UP 36.50

Probability and its Applications

Gregory F. Lawler

Intersections of Random Walks

Birkhäuser
Boston • Basel • Berlin

Gregory F. Lawler
Department of Mathematics
Duke University
Durham, NC 27706
U.S.A.

Library of Congress Cataloging-in-Publication Data

Lawler, Gregory F., 1955-
    Intersections of random walks / Gregory F. Lawler.
    p. cm. -- (Probability and its applications)
    Includes bibliographical references (p.   -) and index.
    ISBN 0-8176-3892-X (soft : alk. paper) ISBN 3-7643-3892-X
    (soft: alk. paper)
    1. Random walks (Mathematics) I. Title. II. Series.
QA274.73.L38 1996   96-17863
    519.2'82--dc20   CIP

Printed on acid-free paper

© 1996 Birkhäuser Boston   *Birkhäuser* ®

Hardcover edition, printed in 1991 by Birkhäuser Boston

ISBN 0-8176-3892-X
ISBN 3-7643-3892-X

Camera-ready text prepared in L^AT_EX by the author.
Printed and bound by Quinn-Woodbine, Woodbine, NJ.
Printed in the U.S.A.
9 8 7 6 5 4 3 2 1

Preface

A more accurate title for this book would be "Problems dealing with the non-intersection of paths of random walks." These include: harmonic measure, which can be considered as a problem of nonintersection of a random walk with a fixed set; the probability that the paths of independent random walks do not intersect; and self-avoiding walks, i.e., random walks which have no self-intersections. The prerequisite is a standard measure theoretic course in probability including martingales and Brownian motion.

The first chapter develops the facts about simple random walk that will be needed. The discussion is self-contained although some previous exposure to random walks would be helpful. Many of the results are standard, and I have borrowed from a number of sources, especially the excellent book of Spitzer [65]. For the sake of simplicity I have restricted the discussion to simple random walk. Of course, many of the results hold equally well for more general walks. For example, the local central limit theorem can be proved for any random walk whose increments have mean zero and finite variance. Some of the later results, especially in Section 1.7, have not been proved for very general classes of walks. The proofs here rely heavily on the fact that the increments of simple random walk are bounded and symmetric. While the proofs could be easily adapted for other random walks with bounded and symmetric increments, it is not clear how to extend them to more general walks. Some progress in this direction has been made in [59].

The proof of the local central limit theorem in Section 1.2 follows closely the proof in [65]. The next sections develop the usual probabilistic tools for analyzing walks: stopping times, the strong Markov property, martingales derived from random walks, and boundary value problems for discrete harmonic functions. Again, all of this material is standard. The asymptotics of the Green's function for $d \geq 3$ and of the potential kernel for $d = 2$ are then derived. There is care in these sections in being explicit about the size of the error in asymptotic results. While this makes it a little harder to read initially, it is hoped that this will allow the chapter to be a reference for "well known" facts about simple random walks. The results in the last section of this chapter are analogous to results which are standard in partial differential equations: difference estimates for harmonic functions and Harnack inequality. Unfortunately the discrete versions of these useful results do not seem to be familiar to many people working in random walks. A version of Theorem 1.7.1(a) was first proved in [8]. A number of "exercises" are included in Chapter 1 and the beginning of Chapter 2. It is suggested that the reader do the exercises, and I have felt free to quote results from the exercises later in the book.

Harmonic measure is the subject of the second chapter. By harmonic measure here we mean harmonic measure from infinity, i.e., the hitting distribution of a set from a random walker starting at infinity. There are many ways to show the existence of harmonic measure, see e.g. [65]. Here the existence is derived as a consequence of the results in Section 1.7. This method has the advantage that it gives a bound on the rate of convergence. In Sections 2.2 and 2.3, the idea of discrete capacity is developed. The results of these sections are well known although some of the proofs are new. I take the viewpoint here that capacity is a measure of the probability that a random walk will hit a set. In the process, I completely ignore the interpretation in terms of electrical capacity or equilibrium potentials. Computing harmonic measure or escape probabilities can be very difficult. Section 2.4 studies the example of a line or a line segment and in the process develops some useful techniques for estimating harmonic measure. First, there is a discussion of Tauberian theorems which are used to relate random walks with geometric killing times with random walks with a fixed number of steps (analytically, this is a comparison of a sequence and its generating function). Then the harmonic measure of a line and a line segment are derived. The earlier estimates are standard. The estimate for the endpoint of a line segment in two dimensions (2.41) was first derived by Kesten [35] using a different argument. The argument here which works for two and three dimensions first appeared in [45]. The next section gives upper bounds for harmonic measure. The bound in terms of the cardinality of the set has been known for a long time. The bound for connected sets in terms of the radius is a discrete analogue of the Beurling projection theorem (see [1]) and was first proved for $d = 2$ by Kesten [35]. The three dimensional result is new here; however, the proofs closely follow those in [35]. The final section gives an introduction to diffusion limited aggregation (DLA), a growth model first introduced by Witten and Sander [73]. The bounds from the previous section are used to give bounds on the growth rate of DLA clusters; again, the result for $d = 2$ was first proved by Kesten [36] and the three dimensional result uses a similar proof.

The next three chapters study the problem of intersections of random walks or, more precisely, the probability that the paths of independent random walks intersect. We will not discuss in detail what the typical intersection set looks like. This has been studied by a number of authors under the name "intersection local time", see e.g. [47] . The discussion on the probability of intersection follows the results in [11, 12, 39, 40, 41, 45]. Chapter 3 sets the basic framework and proves some of the easier results. In Section 3.2, the expected number of intersections is calculated (a straightforward computation) and one lower bound on the hitting probability is given, using a proof adapted from [22]. The expected number of intersections gives a natural conjecture about the order of the probability of "long-range" intersections. This conjecture is proved in the next two sections. For $d \neq 4$, the proof requires little more than the estimate of two moments of the number of intersections. More work is needed in the critical dimension $d = 4$; the proof we give in Section 3.4 uses the properties of a certain random variable which has a small variance in four dimensions. This random variable is used in the next chapter when more precise estimates are given for $d = 4$. The problem of estimating the probability of intersections of two random walks starting at the same point is then considered. It turns out that the easier problem to discuss is the probability that a "two-sided" walk does not intersect a "one-sided" walk. The probability of no intersection in this case is shown to be equal to the inverse of the expected number of intersections, at least up to a multiplicative constant. This fact is proved in Sections 3.5, 3.6, and 4.2. This then gives some upper and lower bounds for the probability that two one-sided walks starting at the same point do not intersect. The material in this chapter essentially follows the arguments in [39, 45]. Some of these results have been obtained by other means [2, 23, 58], and some simplifications from those papers are reflected in the treatment here .

The techniques of Chapter 3 are not powerful enough to analyze the probability that two one-sided walks starting at the origin do not intersect. There are a number of reasons to be interested in this problem. It is a random walk analogue of a quantity that arises in a number of problems in mathematical physics (e.g., a similar quantity arises in the discussion of critical exponents for self-avoiding walks in Section 6.3). Also, some of the techniques used in nonrigorous calculations in mathematical physics can be applied to this problem, see e.g. [16, 17], so rigorous analysis of this problem can be used as a test of the effectiveness of these nonrigorous methods. Unfortunately, there is not yet a complete solution to this problem; Chapters 4 and 5 discuss what can be proved.

In four dimensions, the probability of nonintersection goes to zero like an inverse power of the logarithm of the number of steps. The techniques of Chapter 3 give bounds on this power; in Chapter 4, the exact power is derived. The first part of the derivation is to give asymptotic expressions for the probability of "long-range" intersections (the results of the previous chapter only give expressions up to a multiplicative constant). Sections 4.3 and 4.4 derive the expressions, using a natural relationship between long-range intersections and intersections of a two-sided walk with a one-sided walk. The next section derives the exact power of the logarithm. It essentially combines the result on long-range intersection with an estimate on asymptotic independence of short-range and long-range intersections to estimate the "derivative" of the probability of no intersection. The final section discusses a similar problem, the mutual intersections of three walks in three dimensions. The results are analogous to those of two walks in four dimensions. Some of these results appeared in [41]. One new result is Theorem 4.5.4, which gives the exact power of the logarithm for the probability of no intersection.

The next chapter considers the intersection probability in dimensions two and three. Here the probability of no intersection goes to zero like a power of the number of steps. Again, the results of Chapter 3 can be used to give upper and lower bounds for the exponent. The first thing that is proved is that the exponent exists. This is done in Sections 5.2 and 5.3 by relating it to an exponent for intersections of paths of Brownian motions. Some estimates are derived for the exponent in the remainder of the chapter. First a variational formulation of the exponent is given. The formulation is in terms of a function of Brownian motion. Bounds on this function then give bounds on the exponent. Section 5.5 gives a lower bound for the intersection exponent in two dimensions by comparing it to a different exponent which measures the probability that a Brownian motion makes a closed loop around the origin. The last section gives an upper bound in two and three dimensions.

The last two chapters are devoted to self-avoiding walks, i.e., random walks conditioned to have no (or few) self-intersections. Sections 6.2 and 6.3 discuss the usual (strictly) self-avoiding walk, i.e., simple random walk of a given length with no self-intersections. The connective constant is defined, and then there is a discussion of the critical exponents for the model. The critical exponents are discussed from a probabilistic viewpoint; however, the discussion is almost entirely heuristic. The few nontrivial results about the self-avoiding walk have been obtained from either combinatorial or (mathematical physics) field-theoretic arguments. We mention a few of these results here. There is a forthcoming book by N. Madras and G. Slade in this series which will cover these topics in more detail. The next two sections discuss other models for self-avoiding or self-repelling walks. They fall neatly into two categories: configurational models (Section 6.4) and kinetically growing walks (Section 6.5). The final section gives a brief introduction to the problem of producing self-avoiding walks on the computer, a topic which has raised a number of interesting mathematical questions.

The last chapter discusses a particular model for self-avoiding walks, the loop-erased or Laplacian random walk. This model can be defined in two equivalent ways, one by erasing loops from the paths of simple random walk and the other as a kinetically growing walks with steps taken weighted according to harmonic measure. This model is similar to the usual self-avoiding walk in a number of ways: the critical dimension is four; there is convergence to Brownian motion for dimensions greater than or equal to four, with a logarithmic correction in four dimensions; nontrivial exponents describe the mean-square displacement below four dimensions. Unfortunately, this walk is not in the same universality class as the usual self-avoiding walk; in particular, the mean-square displacement exponent is different. The basic construction of the process is done in the first four sections. There are some technical difficulties in defining the walk in two dimensions because of the recurrence of simple random walk. These are discussed in Section 7.4. In the next section, estimates on the average amount erased are made. These are then used in Section 7.6 to show that the mean-square displacement exponents are at least as large as the Flory exponents for usual self-avoiding walk. The convergence to Brownian motion in high dimensions is done in the last section. Essentially the result follows from a weak law that says that the amount erased is uniform on each path. The proof follows [38, 42]; however, unlike those papers the treatment in this book does not use any nonstandard analysis.

A number of people have made useful comments during the preparation of this book. I would especially like to thank Tom Polaski and Harry Kesten. Partial support for this work was provided by the National Science Foundation, the Alfred P. Sloan Research Foundation, and the U.S. Army Research Office through the Mathematical Sciences Institute at Cornell University.

In the second printing of this book, a number of misprints have been corrected. I would like to thank Ted Sweet and Sungchul Lee for sending me misprints that they found. I have also added a short addendum that updates the status of some of the problems mentioned in the last four chapters of the book.

Notation

We use c, c_1, c_2 to denote arbitrary positive constants, depending only on dimension, which may change from line to line. If a constant is to depend on some other quantity, this will be made explicit. For example, if c depends on α, we write $c(\alpha)$ or c_α. If $g(x), h(x)$ are functions we write $g \sim h$ if they are asymptotic, i.e,

$$\lim_{x \to \infty} \frac{h(x)}{g(x)} = 1.$$

We write $g \asymp h$ if there exist constants c_1, c_2 such that

$$c_1 g(x) \leq h(x) \leq c_2 g(x).$$

Finally we write $g \approx h$ if $\ln g \sim \ln h$.

We write $h(x) = O(g(x))$ if $h(x) \leq cg(x)$ for some constant c. Again, the implicit assumption is that the constant c depends only on dimension. If we wish to imply that the constant may depend on another quantity, say α, we write $O_\alpha(g(x))$. For example, $\alpha x = O_\alpha(x)$, but it is not true that $\alpha x = O(x)$. Similarly, we write $h(x) = o(g(x))$ if $h(x)/g(x) \to 0$. By implication, the rate of convergence depends on no other parameters, except dimension. We will write o_α to indicate a dependence on the parameter α.

Similar conventions hold for limits as $x \to 0$ or $x \to 1-$.

Contents

Chapter 1

Simple Random Walk

1.1 Introduction

Let $X_1, X_2, \ldots$ be independent, identically distributed random variables defined on a probability space $(\Omega, \mathcal{F}, P)$ taking values in the integer lattice Z^d with

$$P\{X_j = e\} = \frac{1}{2d}, \quad |e| = 1.$$

A *simple random walk* starting at $x \in Z^d$ is a stochastic process S_n, indexed by the nonnegative integers, with $S_0 = x$ and

$$S_n = x + X_1 + \cdots + X_n.$$

The probability distribution of S_n is denoted by

$$p_n(x, y) = P^x\{S_n = y\}.$$

Here we have written P^x to indicate that the simple random walk starts at the point x. We will similarly write E^x to denote expectations assuming $S_0 = x$. If the x is missing, it will be assumed that $S_0 = 0$. We let $p_n(x) = p_n(0, x)$. We will sometimes write $p(n, x)$ for $p_n(x)$. It follows immediately that the following hold:

$$
\begin{aligned}
p_n(x, y) &= p_n(y, x), & (1.1) \\
p_n(x, y) &= p_n(y - x), & (1.2) \\
p_n(x) &= p_n(-x), & (1.3) \\
p_0(x, y) &= \delta(y - x), & (1.4)
\end{aligned}
$$

11

where δ is the standard delta function, $\delta(0) = 1, \delta(x) = 0$ if $x \neq 0$. If m is any positive integer, then the process

$$
\begin{aligned}
\tilde{S}_n &= S_{n+m} - S_m \\
&= X_{m+1} + \cdots + X_{m+n}
\end{aligned}
$$

is a simple random walk starting at 0, independent of $\{X_1, \cdots, X_m\}$. From this we can derive

$$
p_{m+n}(x, y) = \sum_{z \in Z^d} p_m(x, z) p_n(z, y). \tag{1.5}
$$

1.2 Local Central Limit Theorem

What is the behavior of $p_n(x)$ for large n? Assume $S_0 = 0$. Then S_n is a sum of independent random variables with mean 0 and covariance $\frac{1}{d} I$. The central limit theorem states that $n^{-1/2} S_n$ converges in distribution to a normal random variable in R^d with mean 0 and covariance $\frac{1}{d} I$, i.e., if $A \subset R^d$ is an open ball,

$$
\lim_{n \to \infty} P\{ \frac{S_n}{\sqrt{n}} \in A \} = \int_A (\frac{d}{2\pi})^{d/2} e^{-\frac{d|x|^2}{2}} \, dx_1 dx_2 \cdots dx_d.
$$

Of course, the random variable S_n only takes on values in Z^d. Moreover, if n is even, then S_n has even parity, i.e., the sum of its components is even, while S_n has odd parity for n odd. A typical open ball $A \subset R^d$ contains about $n^{d/2} m(A)$ points in the lattice $n^{-1/2} Z^d$, where m denotes Lebesgue measure. About half of these points will have even parity. Therefore if n is even, and the random walk spreads itself as evenly as possible among the possible lattice points we might expect

$$
P\{ \frac{S_n}{\sqrt{n}} = \frac{x}{\sqrt{n}} \} \approx \frac{2}{n^{d/2}} (\frac{d}{2\pi})^{d/2} e^{-\frac{d|x|^2}{2n}}.
$$

The local central limit theorem makes this statement precise.

The proof of the local central limit theorem, like the standard proof of the usual central limit theorem, consists of analysis of the characteristic function for simple random walk. If Y is any random variable taking values in Z^d, the characteristic function $\phi(\theta) = \phi_Y(\theta), \theta = (\theta_1, \ldots, \theta_d)$, given by

$$
\phi(\theta) = E(e^{iY \cdot \theta}) = \sum_{y \in Z^d} P\{Y = y\} e^{iy \cdot \theta}, \tag{1.6}
$$

has period 2π in each component. We can therefore think of ϕ as a function on $[-\pi, \pi]^d$ with periodic boundary conditions. The inversion formula for the characteristic function is

$$P\{Y = y\} = \frac{1}{(2\pi)^d} \int_{[-\pi,\pi]^d} e^{-iy\cdot\theta} \phi(\theta) d\theta. \tag{1.7}$$

This can be derived from (1.6) by multiplying both sides by $e^{-iy\cdot\theta}$, integrating, and noting that

$$\int_{[-\pi,\pi]^d} e^{ix\cdot\theta} d\theta = (2\pi)^d \delta(x).$$

The characteristic function ϕ^n for S_n can be computed easily, $\phi^n(\theta) = [\phi(\theta)]^n$, where

$$\phi(\theta) = \frac{1}{d} \sum_{j=1}^{d} \cos\theta_j.$$

We will now prove a very strong version of the local central limit theorem which will be useful throughout this book. Let $\overline{p}_0(x) = \delta(x)$ and for $n > 0$,

$$\overline{p}_n(x) = \overline{p}(n, x) = 2(\frac{d}{2\pi n})^{d/2} e^{-\frac{d|x|^2}{2n}}.$$

We write $n \leftrightarrow x$ if n and x have the same parity, i.e., if $n + x_1 + \cdots + x_d$ is even. Similarly we will write $x \leftrightarrow y$ and $n \leftrightarrow m$. We define the error $E(n, x)$ by

$$E(n, x) = \begin{cases} p(n, x) - \overline{p}(n, x) & \text{if } n \leftrightarrow x, \\ 0 & \text{if } n \nleftrightarrow x. \end{cases}$$

If $f : Z^d \to R$ is any function and $y \in Z^d$, we let $\nabla_y f$ and $\nabla_y^2 f$ be the first and second differences in the direction y defined by

$$\begin{aligned} \nabla_y f(x) &= f(x + y) - f(x), \\ \nabla_y^2 f(x) &= f(x + y) + f(x - y) - 2f(x). \end{aligned}$$

If $f : R^d \to R$ is a C^3 function, $x, y \in Z^d$, $y = |y|u$, then Taylor's theorem with remainder gives

$$|\nabla_y f(x) - |y| D_u f(x)| \leq \frac{1}{2}|y|^2 \sup_{0 \leq a \leq 1} |D_{uu} f(x + ay)| \tag{1.8}$$

$$|\nabla_y^2 f(x) - |y|^2 D_{uu} f(x)| \leq \frac{1}{3}|y|^3 \sup_{0 \leq a \leq 1} |D_{uuu} f(x + ay)|. \tag{1.9}$$

Theorem 1.2.1 (Local Central Limit Theorem) *If $E(n,x)$ is defined as above, then*

$$|E(n,x)| \leq O(n^{-(d+2)/2}), \tag{1.10}$$
$$|E(n,x)| \leq |x|^{-2}O(n^{-d/2}). \tag{1.11}$$

Moreover, if $y \leftrightarrow 0$, there exists a $c_y < \infty$ such that

$$|\nabla_y E(n,x)| \leq c_y O(n^{-(d+3)/2}), \tag{1.12}$$
$$|\nabla_y E(n,x)| \leq c_y |x|^{-2}O(n^{-(d+1)/2}), \tag{1.13}$$
$$|\nabla_y^2 E(n,x)| \leq c_y O(n^{-(d+4)/2}), \tag{1.14}$$
$$|\nabla_y^2 E(n,x)| \leq c_y |x|^{-2}O(n^{-(d+2)/2}). \tag{1.15}$$

Proof. We may assume $n \leftrightarrow x$. By (1.7),

$$p_n(x) = (2\pi)^{-d} \int_{[-\pi,\pi]^d} e^{-ix\cdot\theta} \phi^n(\theta)d\theta.$$

Since $n \leftrightarrow x$, the integrand is not changed if we replace θ with $\theta + (\pi,\dots,\pi)$. Therefore

$$p_n(x) = 2(2\pi)^{-d} \int_A e^{-ix\cdot\theta} \phi^n(\theta)d\theta,$$

where $A = [-\pi/2, \pi/2] \times [-\pi, \pi]^{d-1}$. From the series expansion about 0 of ϕ, $\phi(\theta) = 1 - \frac{1}{2d}|\theta|^2 + O(|\theta|^4)$, we can find an $r \in (0, \pi/2)$ such that $\phi(\theta) \leq 1 - \frac{1}{4d}|\theta|^2$ for $|\theta| \leq r$. There exists a $\rho < 1$, depending on r, such that $|\phi(\theta)| \leq \rho$ for $\theta \in A$, $|\theta| \geq r$. Hence $p(n,x) = I(n,x) + J(n,x)$ where

$$I(n,x) = 2(2\pi)^{-d} \int_{|\theta|\leq r} e^{-ix\cdot\theta} \phi^n(\theta)d\theta,$$

and $|J(n,x)| \leq \rho^n$. If we let $\alpha = \sqrt{n}\theta$,

$$I(n,x) = 2(2\pi\sqrt{n})^{-d} \int_{|\alpha|\leq r\sqrt{n}} \exp\{-\frac{ix\cdot\alpha}{\sqrt{n}}\}\phi^n(\frac{\alpha}{\sqrt{n}})d\alpha.$$

We decompose $I(n,x)$ as follows:

$$\frac{1}{2}(2\pi\sqrt{n})^d I(n,x) = I_0(n,x) + I_1(n,x) + I_2(n,x) + I_3(n,x),$$

where

$$I_0(n,x) = \int_{R^d} \exp\{-\frac{ix\cdot\alpha}{\sqrt{n}}\}\exp\{-\frac{|\alpha|^2}{2d}\}d\alpha,$$

$$I_1(n, x) = \int_{|\alpha| \le n^{1/4}} [\phi^n(\frac{\alpha}{\sqrt{n}}) - \exp\{-\frac{|\alpha|^2}{2d}\}] \exp\{-\frac{ix \cdot \alpha}{\sqrt{n}}\} d\alpha,$$

$$I_2(n, x) = -\int_{|\alpha| \ge n^{1/4}} \exp\{-\frac{|\alpha|^2}{2d}\} \exp\{-\frac{ix \cdot \alpha}{\sqrt{n}}\} d\alpha,$$

$$I_3(n, x) = \int_{n^{1/4} \le |\alpha| \le rn^{1/2}} \phi^n(\frac{\alpha}{\sqrt{n}}) \exp\{-\frac{ix \cdot \alpha}{\sqrt{n}}\} d\alpha.$$

The first integral can be computed exactly by completing the square in the exponential,

$$I_0(n, x) = (2\pi d)^{d/2} e^{-\frac{d|x|^2}{2n}}.$$

Therefore,

$$E(n, x) = J(n, x) + 2(2\pi\sqrt{n})^{-d} \sum_{j=1}^{3} I_j(n, x).$$

We will now bound each of the above terms uniformly in x. We have already noted that $|J(n, x)| \le \rho^n$.

$$|I_1(n, x)| \le \int_{|\alpha| \le n^{1/4}} |\phi^n(\frac{\alpha}{\sqrt{n}}) - \exp\{-\frac{|\alpha|^2}{2d}\}| d\alpha.$$

For $|\alpha| \le n^{1/4}$,

$$\phi(\frac{\alpha}{\sqrt{n}}) = 1 - \frac{|\alpha|^2}{2dn} + |\alpha|^4 O(n^{-2}),$$

$$\phi^n(\frac{\alpha}{\sqrt{n}}) = (1 - \frac{|\alpha|^2}{2dn} + |\alpha|^4 O(n^{-2}))^n$$

$$= (1 - \frac{|\alpha|^2}{2dn})^n (1 + |\alpha|^4 O(n^{-1}))$$

$$= \exp\{-\frac{|\alpha|^2}{2d}\}(1 + |\alpha|^8 O(n^{-1})).$$

Therefore,

$$|I_1(n, x)| \le O(n^{-1}) \int_{|\alpha| \le n^{1/4}} |\alpha|^8 \exp\{-\frac{|\alpha|^2}{2d}\} d\alpha$$

$$= O(n^{-1}).$$

$$|I_2(n, x)| \le \int_{|\alpha| \ge n^{1/4}} \exp\{-\frac{|\alpha|^2}{2d}\} d\alpha$$

$$= c \int_{n^{1/4}}^{\infty} r^{d-1} \exp\{-\frac{r^2}{2d}\} dr$$

$$= O(n^{(d-1)/4} \exp\{-\frac{n^{1/2}}{2d}\}).$$

$$|I_3(n,x)| \leq \int_{n^{1/4} \leq |\alpha| \leq rn^{1/2}} \phi^n(\frac{\alpha}{\sqrt{n}}) d\alpha$$

$$\leq \int_{n^{1/4} \leq |\alpha| \leq rn^{1/2}} (1 - \frac{|\alpha|^2}{4dn})^n$$

$$\leq \int_{n^{1/4} \leq |\alpha|} \exp\{-\frac{|\alpha|^2}{4d}\} d\alpha$$

$$= O(n^{(d-1)/4} \exp\{-\frac{n^{1/2}}{4d}\}).$$

This proves (1.10).

We now consider (1.12) and (1.14).

$$|\nabla_y E(n,x)| \leq |\nabla_y J(n,x)| + \sum_{j=1}^{3} |\nabla_y I_j(n,x)|.$$

Each of these terms except for $\nabla_y I_1$ can be estimated easily by adding the appropriate terms, e.g.,

$$|\nabla_y I_2(n,x)| \leq |I_2(n,x)| + |I_2(n,x+y)|$$

$$\leq O(n^{(d-1)/4} \exp\{-\frac{n^{1/2}}{2d}\})$$

$$= o(n^{-3/2}).$$

The I_1 term, which is the largest of the error terms, requires a little more care:

$$|\nabla_y I_1(n,x)| \leq \int_{|\alpha| \leq n^{1/4}} |1 - \exp\{-\frac{y \cdot \alpha}{\sqrt{n}}\}| |\phi^n(\frac{\alpha}{\sqrt{n}}) - \exp\{-\frac{|\alpha|^2}{2d}\}| d\alpha$$

$$\leq \int_{|\alpha| \leq n^{1/4}} (\frac{|y||\alpha|}{\sqrt{n}})(|\alpha|^8 \exp\{-\frac{|\alpha|^2}{2d}\} O(n^{-1})) d\alpha$$

$$= |y| O(n^{-3/2}) \int_{|\alpha| \leq n^{1/4}} |\alpha|^9 \exp\{-\frac{|\alpha|^2}{2d}\} d\alpha$$

$$= |y| O(n^{-3/2}).$$

This gives (1.12). To get (1.14), again we write

$$|\nabla_y^2 E(n,x)| \leq |\nabla_y^2 J(n,x)| + \sum_{j=1}^{3} |\nabla_y^2 I_j(n,x)|.$$

Again each term other than the I_1 term can be estimated easily. For that term we get,

$$
\begin{aligned}
|\nabla_y^2 I_1(n,x)| &\leq \int_{|\alpha|\leq n^{1/4}} |2 - \exp\{-\frac{y\cdot\alpha}{\sqrt{n}}\} - \exp\{\frac{y\cdot\alpha}{\sqrt{n}}\}| \\
&\qquad |\phi^n(\frac{\alpha}{\sqrt{n}}) - \exp\{-\frac{|\alpha|^2}{2d}\}|d\alpha \\
&= \int_{|\alpha|\leq n^{1/4}} [c_y|\alpha|^2 O(n^{-1})][|\alpha|^8 \exp\{-\frac{|\alpha|^2}{2d}\}O(n^{-1})]d\alpha \\
&\leq c_y O(n^{-2}) \int_{|\alpha|\leq n^{1/4}} |\alpha|^{10} \exp\{-\frac{|\alpha|^2}{2d}\}d\alpha \\
&= c_y O(n^{-2}).
\end{aligned}
$$

To get (1.11),(1.13), and (1.15) we use Green's formula on (1.7),

$$
\begin{aligned}
|x|^2 p_n(x) &= \int_{[-\pi,\pi]^d} |x|^2 e^{-ix\cdot\theta}\phi^n(\theta)d\theta \\
&= \int_{[-\pi,\pi]^d} (-\Delta e^{-ix\cdot\theta})\phi^n(\theta)d\theta \\
&= \int_{[-\pi,\pi]^d} e^{-ix\cdot\theta}(-\Delta\phi^n(\theta))d\theta.
\end{aligned}
$$

(The boundary terms disappear by periodicity.) A direct calculation gives

$$
\Delta\phi^n(\theta) = n(n-1)\phi^{n-2}(\theta)(\frac{1}{d^2}\sum_{j=1}^{d}\sin^2\theta_j) - n\phi^n(\theta).
$$

The proof then proceeds as above for (1.10), (1.12), (1.14), splitting the integral into similar pieces. □

Exercise 1.2.2 *Complete the details of Theorem 1.2.1 for (1.11), (1.13), and (1.15).*

It follows from Theorem 1.2.1 that simple random walk eventually "forgets its starting point."

Corollary 1.2.3 *Suppose $x \leftrightarrow y$. Then*

$$
\lim_{n\to\infty} \sum_{z\in Z^d} |p_n(x,z) - p_n(y,z)| = 0. \tag{1.16}
$$

Proof. We may assume $x = 0$. By the central limit theorem, for every $\gamma > \frac{1}{2}$,

$$\lim_{n\to\infty} \sum_{|z|\geq n^\gamma} (p_n(0, z) + p_n(y, z)) = 0.$$

Therefore it suffices to prove for some $\gamma > \frac{1}{2}$,

$$\lim_{n\to\infty} \sum_{|z|\leq n^\gamma} |p_n(0, z) - p_n(y, z)| = 0.$$

By the definition of $\bar{p}$ and (1.12),

$$|p_n(z) - p_n(z - y)| \leq |\bar{p}_n(z) - \bar{p}_n(z - y)| + |E(n, z) - E(n, z - y)|$$
$$\leq c_y O(n^{-(d+2)/2}) + c_y O(n^{-(d+3)/2}).$$

Therefore,

$$\sum_{|z|\leq n^\gamma} |p_n(z) - p_n(z - y)| \leq \sum_{z\leq n^\gamma} c_y O(n^{-(d+2)/2})$$
$$= c_y O(n^{(2d\gamma-d-2)/2}),$$

which goes to zero if $\gamma < \frac{1}{2} + \frac{1}{d}$. $\square$

Exercise 1.2.4 *Prove for every $m \leftrightarrow 0$,*

$$\lim_{n\to\infty} \sum_{z\in Z^d} |p_n(z) - p_{n+m}(z)| = 0. \qquad (1.17)$$

(Hint: Use (1.5) and Corollary 1.2.3.)

There is another approach to the local central limit theorem for simple random walk. Suppose $d = 1$, and x is a positive integer. Then by the binomial distribution,

$$P\{S_{2n} = 2x\} = \frac{(2n)!}{(n + x)!(n - x)!}(\frac{1}{2})^{2n}.$$

We can estimate this expression using Stirling's formula [24, (9.15)],

$$n! = \sqrt{2\pi} n^{n+\frac{1}{2}} e^{-n}(1 + O(\frac{1}{n})).$$

If $\alpha < 2/3$, we can plug in, do some calculation, and get for $|x| \leq n^\alpha$,

$$P\{S_{2n} = 2x\} = (\pi n)^{-1/2} \exp\{-\frac{(2x)^2}{4n}\}(1 + O(n^{3\alpha-2}))$$
$$= \bar{p}(2n, 2x)(1 + O(n^{3\alpha-2})). \qquad (1.18)$$

This statement is not as strong as Theorem 1.2.1 when $|x|$ is of order $\sqrt{n}$ but gives more information when $|x|$ is significantly larger than $\sqrt{n}$. A similar argument can be done for $d > 1$ (although it is messier to write down) which we will omit. However, we state the result for future reference.

Proposition 1.2.5 *If* $\alpha < 2/3$, *then if* $|x| \leq n^{\alpha}, 0 \leftrightarrow x \leftrightarrow n$,

$$p(n, x) = \bar{p}(n, x)(1 + O(n^{3\alpha - 2})).$$

1.3 Strong Markov Property

A *random time* is any random variable $\tau : \Omega \rightarrow \{0, 1, 2, \ldots\} \cup \{\infty\}$. A stopping time for a random walk is any random time which depends only on the "past and present." To formalize this idea let $\mathcal{F}_n$ and $\mathcal{H}_n$ be the σ-algebras of the "past and present" and the "future" respectively, i.e.,

$$
\begin{aligned}
\mathcal{F}_n &= \sigma\{X_1, \ldots, X_n\}, \\
\mathcal{H}_n &= \sigma\{X_{n+1}, X_{n+2}, \ldots\}.
\end{aligned}
$$

Then $\mathcal{F}_n$ and $\mathcal{H}_n$ are independent σ-algebras (written $\mathcal{F}_n \perp \mathcal{H}_n$). We will call an increasing sequence of σ-algebras $\mathcal{G}_0 \subset \mathcal{G}_1 \subset \mathcal{G}_2 \subset \cdots$ a *filtration* for the simple random walk if for each n, $\mathcal{F}_n \subset \mathcal{G}_n$ and $\mathcal{G}_n \perp \mathcal{H}_n$. A random time τ is called a *stopping time* (with respect to $\mathcal{G}_n$) if for each $n < \infty$,

$$\{\tau = n\} \in \mathcal{G}_n.$$

Examples

1. If A is any subset of Z^d and k is any integer, then

$$\tau = \inf\{n \geq k : S_n \in A\}$$

is a stopping time with respect to $\mathcal{F}_n$. Most of the stopping times we will consider in this book will be of this form.

2. If τ_1 and τ_2 are stopping times, then $\tau_1 \vee \tau_2$ and $\tau_1 \wedge \tau_2$ are stopping times.

3. Let $Y_0, Y_1, \ldots$ be independent random variables which are independent of $\{X_1, X_2, \ldots\}$ with $P\{Y_i = 1\} = 1 - P\{Y_i = 0\} = \lambda$, and let

$$T = \inf\{j \geq 0 : Y_j = 1\}.$$

We think of T as a "killing time" for the random walk with rate λ. T has a geometric distribution

$$P\{T = j\} = (1 - \lambda)^j \lambda.$$

Let $\mathcal{G}_n = \sigma\{X_1, \ldots, X_n, Y_0, \ldots, Y_n\}$. Then $\mathcal{G}_n$ is a filtration for simple random walk and T is a stopping time with respect to $\mathcal{G}_n$. This will be the only example of a filtration other than $\mathcal{F}_n$ that we will need in this book.

If τ is a stopping time with respect to $\mathcal{G}_n$, the σ-algebra $\mathcal{G}_\tau$ is the collection of events $A \in \mathcal{F}$ such that for each n,

$$A \cap \{\tau \leq n\} \in \mathcal{G}_n.$$

Exercise 1.3.1 *Show that $\mathcal{G}_\tau$ is a σ-algebra.*

Theorem 1.3.2 (Strong Markov Property) *Suppose τ is a stopping time with respect to a filtration $\mathcal{G}_n$. Then on $\{\tau < \infty\}$ the process*

$$\tilde{S}_n = S_{n+\tau} - S_\tau$$

is a simple random walk independent of $\mathcal{G}_\tau$.

Proof. Let $x_0, \ldots, x_n \in Z^d$ and $A \in \mathcal{G}_\tau$. Then

$$P[\{\tilde{S}_0 = x_0, \ldots, \tilde{S}_n = x_n\} \cap A \cap \{\tau < \infty\}]$$

$$= \sum_{j=0}^{\infty} P[\{\tilde{S}_0 = x_0, \ldots, \tilde{S}_n = x_n\} \cap A \cap \{\tau = j\}] =$$

$$= \sum_{j=0}^{\infty} P[\{S_j - S_j = x_0, \ldots, S_{j+n} - S_j = x_n\} \cap A \cap \{\tau = j\}]$$

$$= \sum_{j=0}^{\infty} P\{S_0 = x_0, \ldots, S_n = x_n\} P(A \cap \{\tau = j\})$$

$$= P\{S_0 = x_0, \ldots, S_n = x_n\} P(A \cap \{\tau < \infty\}) \quad \square$$

As an application of the local central limit theorem and the strong Markov property, we consider the question of recurrence and transience of simple random walk. Let R_n be the number of visits to 0 up through time n, i.e.,

$$R_n = \sum_{j=0}^{n} I\{S_j = 0\},$$

where I denotes the indicator function, and let $R = R_\infty$. By Theorem 1.2.1,

$$E(R_n) = \sum_{j=0}^{n} p_j(0)$$

$$= \sum_{j \le n, j \text{ even}} [2(\frac{d}{2\pi j})^{d/2} + O(j^{-(d+2)/2})]$$

$$\sim \begin{cases} \sqrt{2/\pi}\, n^{1/2} + O(1) & d = 1 \\ \frac{1}{\pi} \ln n + O(1) & d = 2 \\ c + O(n^{(2-d)/2}) & d \ge 3. \end{cases}$$

In particular, $E(R) = \infty$ for $d \le 2$. Let

$$\tau = \inf\{j \ge 1 : S_j = 0\}.$$

Then $R = 1 + \sum_{j=\tau}^{\infty} I\{S_j = 0\}$. By Theorem 1.3.2,

$$E(R) = 1 + P\{\tau < \infty\}E(R),$$

or

$$P\{\tau = \infty\} = \frac{1}{E(R)} \begin{cases} = 0 & \text{if } d \le 2 \\ > 0 & \text{if } d \ge 3. \end{cases} \tag{1.19}$$

Another application of Theorem 1.3.2 shows that if $d \ge 3$,

$$P\{R = j\} = p(1 - p)^{j-1},$$

where $p = P\{\tau = \infty\}$. Summarizing we get,

Theorem 1.3.3 *If $d \le 2$, simple random walk is recurrent, i.e.*

$$P\{S_n = 0 \text{ infinitely often}\} = 1.$$

If $d \ge 3$, simple random walk is transient, i.e.,

$$P\{S_n = 0 \text{ infinitely often}\} = 0.$$

Exercise 1.3.4 (Reflection Principle) *If $a > 0$,*

$$P\{\sup_{1 \le j \le n} |S_j| \ge a\} \le 2P\{|S_n| \ge a\}.$$

(Hint: Consider

$$\tau = \inf\{j : |S_j| \ge a\}.)$$

1.4   Harmonic Functions, Dirichlet Problem

Let e_j be the unit vector in Z^d with j^{th} component 1. If $f : Z^d \to R$, then the *(discrete) Laplacian* of f is defined by

$$\Delta f(x) = [\frac{1}{2d} \sum_{|e|=1} f(x + e)] - f(x)$$

$$= \frac{1}{2d} \sum_{|e|=1} \nabla_e f(x)$$

$$= \frac{1}{2d} \sum_{j=1}^{d} \nabla_{e_j}^2 f(x).$$

The third line resembles the usual definition of the Laplacian of a function on R^d, but the first line is a more natural way to think of the Laplacian – the difference between the mean value of f over the neighbors of x and the value of f at x. The Laplacian is related to simple random walk by

$$\Delta f(x) = E^x[f(S_1) - f(S_0)].$$

We call a function *harmonic (subharmonic, superharmonic)* on A if for each $x \in A$, $\Delta f(x) = 0$ ($\Delta f(x) \geq 0, \Delta f(x) \leq 0$). There is a close relationship between harmonic functions and martingales.

Proposition 1.4.1 *Suppose f is a bounded function, harmonic on A, and*

$$\tau = \inf\{j \geq 0 : S_j \notin A\}.$$

Then $M_n = f(S_{n \wedge \tau})$ is a martingale with respect to $\mathcal{F}_n$.

Proof. Assume $S_0 = x$. By the Markov property,

$$E(f(S_{n+1}) \mid \mathcal{F}_n) = E^{S_n}(f(S_1)) = f(S_n) + \Delta f(S_n).$$

Let $B_n = \{\tau > n\}$. Then $M_{n+1} = M_n$ on B_n^c and

$$
\begin{aligned}
E(M_{n+1} \mid \mathcal{F}_n) &= E(M_{n+1} I_{B_n} \mid \mathcal{F}_n) + E(M_{n+1} I_{B_n^c} \mid \mathcal{F}_n) \\
&= E(f(S_{n+1}) I_{B_n} \mid \mathcal{F}_n) + E(M_n I_{B_n^c} \mid \mathcal{F}_n) \\
&= I_{B_n} E(f(S_{n+1}) \mid \mathcal{F}_n) + M_n I_{B_n^c} \\
&= I_{B_n}(f(S_n) + \Delta f(S_n)) + M_n I_{B_n^c}.
\end{aligned}
$$

But $\Delta f(S_n) = 0$ on B_n. Therefore,

$$
\begin{aligned}
E(M_{n+1} \mid \mathcal{F}_n) &= I_{B_n} f(S_n) + I_{B_n^c} M_n \\
&= M_n. \quad \square
\end{aligned}
$$

Exercise 1.4.2 *Suppose f is a bounded function, superharmonic on A, and $\tau = \inf\{j \geq 0 : S_j \notin A\}$. Show that $M_n = f(S_{n \wedge \tau})$ is a supermartingale with respect to $\mathcal{F}_n$.*

Exercise 1.4.3 *Show that $M_n = |S_n|^2 - n$ is a martingale with respect to $\mathcal{F}_n$.*

A number of results about random walks and harmonic functions can be proved using Proposition 1.4.1 and the optional sampling theorem [9, Theorem 5.10]. For example, let $d = 1$, $f(x) = x$, and

$$\tau = \inf\{j \geq 0 : S_j = 0 \text{ or } S_j = n\}.$$

Then if $0 \leq S_0 \leq n$, $M_n = S_{n \wedge \tau}$ is a bounded martingale and the optional sampling theorem states that for $0 \leq x \leq n$,

$$x = E^x(M_0) = E^x(M_\tau) = nP^x\{S_\tau = n\}.$$

Therefore,

$$P^x\{S_\tau = n\} = \frac{x}{n}. \tag{1.20}$$

Before giving another example we will prove an easy lemma that will be useful later in the book.

Lemma 1.4.4 *If $A \subset Z^d$ is a finite set and*

$$\tau = \inf\{j \geq 1 : S_j \notin A\},$$

then there exist $C < \infty$ and $\rho < 1$ (depending on A) such that for each $x \in A$,

$$P^x\{\tau \geq n\} \leq C\rho^n.$$

Proof. Let $R = \sup\{|x| : x \in A\}$. Then for each $x \in A$, there is a path of length $R + 1$ starting at x and ending outside of A, hence

$$P^x\{\tau \leq R + 1\} \geq (\frac{1}{2d})^{R+1}.$$

By the Markov property,

$$
\begin{aligned}
P^x\{\tau > k(R+1)\} &= P^x\{\tau > (k-1)(R+1)\} \\
&\quad P^x\{\tau > k(R+1) \mid \tau > (k-1)(R+1)\} \\
&\leq P^x\{\tau > (k-1)(R+1)\} (1 - (2d)^{-(R+1)}),
\end{aligned}
$$

and hence

$$P^x\{\tau > k(R+1)\} \leq \rho^{k(R+1)},$$

where $\rho = (1 - (2d)^{-(R+1)})^{1/(R+1)}$. For integer n write $n = k(R+1) + j$ where $j \in \{1, \ldots, R+1\}$. Then

$$
\begin{aligned}
P^x\{\tau \geq n\} &\leq P^x\{\tau > k(R+1)\} \\
&\leq \rho^{k(R+1)} \\
&\leq \rho^{-(R+1)}\rho^n. \quad \square
\end{aligned}
$$

We now consider the martingale $M_n = |S_n|^2 - n$ (Exercise 1.4.3). Let

$$\tau = \inf\{j \geq 1 : |S_j| \geq N\}.$$

By Lemma 1.4.4, if $|x| < N$,

$$E^x(|M_n|I\{\tau \geq n\}) \leq ((N+1)^2 + n)P^x\{\tau \geq n\}$$
$$\longrightarrow 0.$$

We can therefore use the optional sampling theorem to conclude

$$E^x(M_\tau) = E^x(M_0) = |x|^2.$$

But $N^2 - \tau \leq M_\tau < (N+1)^2 - \tau$, and hence

$$N^2 - |x|^2 \leq E^x(\tau) < (N+1)^2 - |x|^2. \tag{1.21}$$

If $A \subset Z^d$, we let

$$\partial A = \{x \notin A : |x - y| = 1 \text{ for some } y \in A\},$$
$$\overline{A} = A \cup \partial A.$$

We are now ready to solve the discrete Dirichlet problem.

Theorem 1.4.5 *Let $A \subset Z^d$ be a finite set and let $F : \partial A \to R$. Then the unique function $f : \overline{A} \to R$ satisfying*

(a) $\Delta f(x) = 0$, $x \in A$,

(b) $f(x) = F(x)$, $x \in \partial A$,

is

$$f(x) = E^x[F(S_\tau)], \tag{1.22}$$

where $\tau = \inf\{j \geq 0 : S_j \notin A\}$.

   Proof. It is easy to check that f defined by (1.22) satisfies (a) and (b). To show uniqueness assume f satisfies (a) and (b) and let $x \in A$. Then $M_n = f(S_{n \wedge \tau})$ is a bounded martingale and by the optional sampling theorem

$$f(x) = E^x(M_0) = E^x(M_\tau) = E^x[F(S_\tau)]. \quad \square$$

   It is not surprising that there is a unique solution to (a) and (b), since (a) and (b) give $|A|$ linear equations in $|A|$ unknowns, where $|\cdot|$ denotes cardinality. The interesting part of the theorem is the nice probabilistic form for the solution. We also get a nice form for the inhomogeneous Dirichlet problem.

Theorem 1.4.6 *Let $A \subset Z^d$ be a finite set, $F : \partial A \to R$, $g : A \to R$. Then the unique function $f : \overline{A} \to R$ satisfying*

$$(a) \ \Delta f(x) = -g(x), \ x \in A,$$

$$(b) \ f(x) = F(x), \ x \in \partial A,$$

is

$$f(x) = E^x[F(S_\tau) + \sum_{j=0}^{\tau-1} g(S_j)]. \tag{1.23}$$

Note that by Lemma 1.4.4 or (1.21),

$$E^x[\sum_{j=0}^{\tau-1} |g(S_j)|] \leq \|g\|_\infty E^x(\tau) < \infty,$$

and so f is well defined.

   Proof. Again it is easy to check that f defined by (1.23) satisfies (a) and (b). To check uniqueness, assume f satisfies (a) and (b), and let M_n be the martingale

$$\begin{aligned} M_n &= f(S_{n\wedge\tau}) - \sum_{j=0}^{(n-1)\wedge(\tau-1)} \Delta f(S_j) \\ &= f(S_{n\wedge\tau}) + \sum_{j=0}^{(n-1)\wedge(\tau-1)} g(S_j). \end{aligned}$$

Note that, by Lemma 1.4.4, $E^x(|M_n|I\{\tau \geq n\}) \leq (\|f\|_\infty + n\|g\|_\infty)P^x\{\tau \geq n\} \to 0$. Therefore by the optional sampling theorem,

$$f(x) = E^x(M_0) = E^x(M_\tau) = E^x[F(S_\tau) + \sum_{j=0}^{\tau-1} g(S_j)]. \quad \square$$

Exercise 1.4.7 (Maximum principle) *If $A \subset Z^d$ is a finite set and $f : \overline{A} \to R$, is subharmonic in A then*

$$\sup_{x \in \overline{A}} f(x) = \sup_{x \in \partial A} f(x).$$

   We now consider the homogeneous Dirichlet problem in the case where A is infinite, e.g., the complement of a finite set. If A is infinite there may be many solutions to the Dirichlet problem. However, we will be able to classify all bounded solutions. Suppose $F : \partial A \to R$ is a bounded function, and as before let

$$\tau = \inf\{j \geq 0 : S_j \in \partial A\}.$$

If $d \leq 2$, then $P^x\{\tau < \infty\} = 1$, and the proof of Theorem 1.4.5 works verbatim to prove the following theorem.

Theorem 1.4.8 *Let $A \subset Z^d, d \leq 2$, and $F : \partial A \to R$ be a bounded function. Then the unique bounded function $f : \overline{A} \to R$ satisfying*

$$\text{(a) } \Delta f(x) = 0, \ x \in A,$$

$$\text{(b) } f(x) = F(x), \ x \in \partial A,$$

is

$$f(x) = E^x[F(S_\tau)].$$

We emphasize that we have proven the existence of a unique <u>bounded</u> solution. It is easy to see that one can have unbounded solutions as well. For example if $d = 1$, $A = Z \backslash \{0\}$, and $F(0) = 0$ then $f(x) = ax$ is a solution to (a) and (b) for any real number a. For $d \geq 3$, Theorem 1.4.8 will hold with the same proof if A is a set with the property that $P^x\{\tau < \infty\} = 1$ for each $x \in A$. This will not be true in general (e.g., if A is the complement of a finite set) because of the transience of the random walk. In fact, if we let

$$f(x) = P^x\{\tau = \infty\}, \tag{1.24}$$

it is easy to check that f is a bounded function satisfying (a) and (b) with $F \equiv 0$. Since $f \equiv 0$ also satisfies (a) and (b) with $F \equiv 0$, we do not have uniqueness. However, the function (1.24) is essentially the only new function that can appear.

Theorem 1.4.9 *Let $A \subset Z^d$ and $F : \partial A \to R^d$ be bounded. Then the only bounded functions $f : \overline{A} \to R$ satisfying*

$$\text{(a) } \Delta f(x) = 0, \ x \in A,$$

$$\text{(b) } f(x) = F(x), x \in \partial A,$$

are of the form

$$f(x) = E^x[F(S_\tau)I\{\tau < \infty\}] + aP^x\{\tau = \infty\}, \tag{1.25}$$

where $a \in R$.

Proof. It is straightforward to check that any f of the form (1.25) satisfies (a) and (b). Suppose that f_1 is a bounded solution to (a) and (b) for a given F and let

$$f(x) = f_1(x) - E^x[F(S_\tau)I\{\tau < \infty\}].$$

Then f is a bounded solution to (a) and (b) with $F \equiv 0$. It suffices to prove that $f(x) = aP^x\{\tau = \infty\}$.

Let $M \leftrightarrow 0$ and let $q_n^M(x, y) = P^x\{S_n = y \mid \tau > M\}$. Since $\{\tau > M\} \in \mathcal{F}_M$, the Markov property implies for $n \geq M$,

$$q_n^M(x, y) = \sum_{z \in Z^d} q_M^M(x, z) p_{n-M}(z, y) \tag{1.26}$$

It follows from (1.16) and (1.17) that for each $M \leftrightarrow 0$, $z \leftrightarrow x$,

$$\lim_{n \to \infty} \sum_{y \in Z^d} |p_n(x, y) - p_{n-M}(z, y)| = 0,$$

and hence for each $M \leftrightarrow 0$,

$$\lim_{n \to \infty} \sum_{y \in Z^d} |p_n(x, y) - q_n^M(x, y)| = 0. \tag{1.27}$$

Since $M_n = f(S_{n \wedge \tau})$ is a martingale, we get for each $n \geq M$,

$$
\begin{aligned}
f(x) &= E^x[M_n] \\
&= E^x[f(S_n)I\{\tau > n\}] \\
&= E^x[f(S_n)I\{\tau > M\}] - E^x[f(S_n)I\{M < \tau \leq n\}] \\
&= P^x\{\tau > M\}E^x[f(S_n) \mid \tau > M] \\
&\quad - E^x[f(S_n)I\{M < \tau \leq n\}] \tag{1.28}
\end{aligned}
$$

The second term is bounded easily,

$$|E^x[f(S_n)I\{M < \tau \leq n\}]| \leq \|f\|_\infty P^x\{M < \tau < \infty\}. \tag{1.29}$$

If $x, z \in A, x \leftrightarrow z$,

$$
\begin{aligned}
&|E^x[f(S_n) \mid \tau > M] - E^z[f(S_n) \mid \tau > M]| \\
&\leq \sum_{y \in Z^d} |f(y)||q_n^M(x, y) - q_n^M(z, y)| \\
&\leq \|f\|_\infty \Big[\sum_{y \in Z^d} |q_n^M(x, y) - p_n(x, y)| \\
&\quad + \sum_{y \in Z^d} |q_n^M(z, y) - p_n(z, y)| + \sum_{y \in Z^d} |p_n(x, y) - p_n(z, y)| \Big].
\end{aligned}
$$

Hence by (1.16) and (1.27),

$$\lim_{n\to\infty} |E^x[f(S_n) \mid \tau > M] - E^z[f(S_n) \mid \tau > M]| = 0.$$

Therefore by (1.28) and (1.29), if $x \leftrightarrow z \leftrightarrow M$,

$$|\frac{f(x)}{P^x\{\tau > M\}} - \frac{f(z)}{P^z\{\tau > M\}}| \le$$
$$\|f\|_\infty[\frac{P^x\{M < \tau < \infty\}}{P^x\{M < \tau\}} + \frac{P^z\{M < \tau < \infty\}}{P^z\{M < \tau\}}].$$

Letting $M \to \infty$, we see if $P^x\{\tau = \infty\} > 0, P^z\{\tau = \infty\} > 0$,

$$f(x)[P^x\{\tau = \infty\}]^{-1} = f(z)[P^z\{\tau = \infty\}]^{-1},$$

i.e., there exists a constant a such that for all $z \leftrightarrow x$,

$$f(z) = aP^z\{\tau = \infty\}.$$

If $y \in A$ with $y \not\leftrightarrow x$, then

$$\begin{aligned}
f(y) &= \frac{1}{2d} \sum_{|e|=1} f(y + e) \\
&= \frac{1}{2d} \sum_{|e|=1} aP^{y+e}\{\tau = \infty\} \\
&= aP^y\{\tau = \infty\}. \quad \square
\end{aligned}$$

Exercise 1.4.10 *Show that any bounded function f which is harmonic on Z^d is constant. (Hint: consider f on $A = Z^d \setminus \{0\}$).*

1.5 Green's Function, Transient Case

If n is a nonnegative integer, we define the *Green's function* $G_n(x, y)$ to be the expected number of visits to y in n steps starting at x, i.e.,

$$\begin{aligned}
G_n(x, y) &= E^x[\sum_{j=0}^{n} I\{S_j = y\}] \\
&= \sum_{j=0}^{n} p_j(x, y) \\
&= \sum_{j=0}^{n} p_j(y - x).
\end{aligned}$$

If $d \geq 3$, we can define $G(x, y) = G_\infty(x, y)$,

$$G(x, y) = \sum_{j=0}^{\infty} p_j(x, y)$$

(if $d \leq 2$, the sum is infinite). We write $G_n(x) = G_n(0, x), G(x) = G(0, x)$. Note that

$$\begin{aligned} \Delta G(x) &= E[\sum_{j=1}^{\infty} I\{S_j = x\}] - E[\sum_{j=0}^{\infty} I\{S_j = x\}] \\ &= E(-I\{S_0 = x\}) = -\delta(x). \end{aligned}$$

The local central limit theorem gives estimates for $p_j(x)$. Here we will use these estimates to study the behavior of the Green's function for large x. As in Theorem 1.2.1, we write $E(n, x) = p(n, x) - \overline{p}(n, x)$ if $n \leftrightarrow x$. If $n \not\leftrightarrow x$, we let $E(n, x) = 0$. As a preliminary, we prove a simple large deviation estimate for simple random walk.

Lemma 1.5.1 *For any $a > 0$, there exists $c_a < \infty$, such that for all $n, t > 0$,*

$$\text{(a) } P\{|S_n| \geq atn^{1/2}\} \leq c_a e^{-t},$$

$$\text{(b) } P\{\sup_{0 \leq i \leq n} |S_i| \geq atn^{1/2}\} \leq 2c_a e^{-t}.$$

Proof. Let $S_n = (S_n^1, \ldots, S_n^d)$. Then,

$$\begin{aligned} P\{|S_n| \geq atn^{1/2}\} &\leq \sum_{j=1}^{d} P\{|S_n^j| \geq d^{-1/2}atn^{1/2}\} \\ &= 2dP\{S_n^1 \geq d^{-1/2}atn^{1/2}\}. \end{aligned}$$

By Chebyshev's inequality,

$$\begin{aligned} P\{S_n^1 \geq d^{-1/2}atn^{1/2}\} &\leq e^{-t}E(\exp\{d^{1/2}a^{-1}n^{-1/2}S_n^1\}) \\ &= e^{-t}[(1 - \frac{1}{d}) + \frac{1}{d}\cosh(\sqrt{d}a^{-1}n^{-1/2})]^n \\ &\leq c_a e^{-t}. \end{aligned}$$

This gives (a), and (b) then follows from the reflection principle (Exercise 1.3.4). $\square$

Lemma 1.5.2 *For every* $\alpha < d$, $y \leftrightarrow 0$,

$$\lim_{|x| \to \infty} |x|^\alpha \sum_{j=0}^\infty |E(j, x)| \;\; = \;\; 0, \tag{1.30}$$

$$\lim_{|x| \to \infty} |x|^{\alpha+1} \sum_{j=0}^\infty |\nabla_y E(j, x)| \;\; = \;\; 0, \tag{1.31}$$

$$\lim_{|x| \to \infty} |x|^{\alpha+2} \sum_{j=0}^\infty |\nabla_y^2 E(j, x)| \;\; = \;\; 0. \tag{1.32}$$

Proof. Let $\gamma < 2$. Then there exists an $a > 0$ such that for $j \leq |x|^\gamma$, $p(j, x) \leq O(\exp\{-|x|^a\})$ (see Lemma 1.5.1) and $\bar{p}(j, x) \leq O(\exp\{-|x|^a\})$, so

$$|E(j, x)| \leq O(\exp\{-|x|^a\}), \; j \leq |x|^\gamma. \tag{1.33}$$

We split the sum into three parts,

$$\sum_{j=0}^\infty |E(j, x)| = \sum_{j \leq |x|^\gamma} + \sum_{|x|^\gamma < j < |x|^2} + \sum_{|x|^2 \leq j < \infty} .$$

By (1.33),

$$\sum_{j \leq |x|^\gamma} |E(j, x)| = O(|x|^\gamma \exp\{-|x|^a\}).$$

By (1.11),

$$\sum_{|x|^\gamma < j < |x|^2} |E(j, x)| \;\; \leq \;\; \sum_{|x|^\gamma < j < |x|^2} |x|^{-2} O(j^{-d/2})$$

$$= \;\; \begin{cases} O(|x|^{-1}), & d = 1, \\ O(|x|^{-2} \ln |x|), & d = 2, \\ O(|x|^{\gamma - \frac{d\gamma}{2} - 2}), & d \geq 3. \end{cases}$$

By (1.10),

$$\sum_{|x|^2 \leq j} |E(j, x)| \;\; \leq \;\; \sum_{|x|^2 \leq j} O(j^{-(d+2)/2})$$

$$= \;\; O(|x|^{-d}).$$

For every $\alpha < d$, by choosing γ sufficiently close to 2 we can then get (1.30). The proofs for (1.31) and (1.32) are similar using (1.12)- (1.15). $\square$

Exercise 1.5.3 *Prove (1.31) and (1.32).*

It will now be easy to give the asymptotics of G for large x. Let $f(x) = |x|^{2-d}$. Note that f is a harmonic function on $R^d \setminus \{0\}$, i.e.,

$$\sum_{j=1}^{d} D_{jj} f(x) = 0.$$

If f is considered as a function on Z^d, by (1.9), as $|x| \to \infty$,

$$\Delta f(x) = O(|x|^{-d-1}), \tag{1.34}$$

where Δ denotes the discrete Laplacian.

Theorem 1.5.4 *If $d \geq 3$, as $|x| \to \infty$,*

$$G(x) \sim a_d |x|^{2-d} \tag{1.35}$$

where

$$a_d = \frac{d}{2} \Gamma(\frac{d}{2} - 1) \pi^{-d/2} = \frac{2}{(d-2)\omega_d},$$

where ω_d is the volume of the unit ball in R^d. Moreover, if $\alpha < d$,

$$\lim_{|x| \to \infty} |x|^\alpha (G(x) - a_d|x|^{2-d}) = 0.$$

Proof: Assume $x \leftrightarrow 0$. By Lemma 1.5.2, for every $\alpha < d$,

$$G(x) = [\sum_{n=0}^{\infty} \bar{p}(2n, x)] + o(|x|^{-\alpha}).$$

But,

$$
\begin{aligned}
\sum_{n=0}^{\infty} \bar{p}(2n, x) &= \sum_{n=0}^{\infty} 2(\frac{d}{4\pi n})^{d/2} e^{-\frac{d|x|^2}{4n}} \\
&= \int_0^\infty 2(\frac{d}{4\pi t})^{d/2} e^{-\frac{d|x|^2}{4t}} dt + O(|x|^{-d}) \\
&= \frac{d}{2} \Gamma(\frac{d}{2} - 1) \pi^{-d/2} |x|^{2-d} + O(|x|^{-d}).
\end{aligned}
$$

If $x \nleftrightarrow 0$, then by (1.34)

$$
\begin{aligned}
G(x) &= \frac{1}{2d} \sum_{|e|=1} G(x+e) \\
&= \frac{1}{2d} \sum_{|e|=1} (a_d|x+e|^{2-d} + o(|x|^{-\alpha})) \\
&= a_d|x|^{2-d} + o(|x|^{-\alpha}). \quad \square
\end{aligned}
$$

We remark that the above result is not quite as strong as can be proved. Note that the dominant term can be estimated

$$\sum_{n=0}^{\infty} \overline{p}(2n, x) = a_d |x|^{2-d} + O(|x|^{-d}).$$

Hence one might guess that

$$G(x) = a_d |x|^{2-d} + O(|x|^{-d}).$$

This is the case, but we will not need this stronger result in this book.

Theorem 1.5.5 If $d \geq 3$, $y \in Z^d$, $y = |y|u$,

$$\nabla_y G(x) - a_d |y| D_u(|x|^{2-d}) = O(|x|^{-d}) \qquad (1.36)$$
$$\nabla_y^2 G(x) - a_d |y|^2 D_{uu}(|x|^{2-d}) = O(|x|^{-d-1}). \qquad (1.37)$$

Proof. First assume $y \leftrightarrow 0, x \leftrightarrow 0$. Then

$$|\nabla_y G(x)| \leq \sum_{n=0}^{\infty} |\nabla_y \overline{p}_{2n}(x)| + \sum_{n=0}^{\infty} |\nabla_y E(n, x)|,$$

$$|\nabla_y^2 G(x)| \leq \sum_{n=0}^{\infty} |\nabla_y^2 \overline{p}_{2n}(x)| + \sum_{n=0}^{\infty} |\nabla_y^2 E(n, x)|.$$

By Lemma 1.5.2,

$$\sum_{n=0}^{\infty} |\nabla_y E(n, x)| = o(|x|^{-d}),$$

$$\sum_{n=0}^{\infty} |\nabla_y^2 E(n, x)| = o(|x|^{-d-1}),$$

so we only need to estimate the dominant term. By (1.8) and (1.9), if $|x| \geq 2|y|$,

$$|\nabla_y e^{-\frac{d|x|^2}{4n}} - |y| D_u e^{-\frac{d|x|^2}{4n}}| \leq c_y e^{-\frac{d|x|^2}{16n}} [|x|^2 O(n^{-2}) + |x| O(n^{-1})],$$

$$|\nabla_y^2 e^{-\frac{d|x|^2}{4n}} - |y|^2 D_{uu} e^{-\frac{d|x|^2}{4n}}| \leq c_y e^{-\frac{d|x|^2}{16n}} [|x|^3 O(n^{-3}) + |x| O(n^{-2})],$$

and therefore,

$$\sum_{n=0}^{\infty} 2(\frac{d}{4\pi n})^{d/2} |\nabla_y e^{-\frac{d|x|^2}{4n}} - |y| D_u e^{-\frac{d|x|^2}{4n}}| \leq c_y O(|x|^{-d}),$$

$$\sum_{n=0}^{\infty} 2(\frac{d}{4\pi n})^{d/2} |\nabla_y^2 e^{-\frac{d|x|^2}{4n}} - |y|^2 D_{uu} e^{-\frac{d|x|^2}{4n}}| \leq c_y O(|x|^{-d-1}).$$

Finally,

$$\sum_{n=0}^{\infty} 2(\frac{d}{4\pi n})^{d/2}|y|D_u e^{-\frac{d|x|^2}{4n}} = \sum_{n=0}^{\infty} 2(\frac{d}{4\pi n})^{d/2}|y|(\frac{dx \cdot u}{2n})e^{-\frac{d|x|^2}{4n}}$$

$$= \int_0^{\infty} 2(\frac{d}{4\pi t})^{d/2}|y|D_u e^{-\frac{d|x|^2}{2t}} dt$$

$$+O(|x|^{-d-1})$$

$$= |y|D_u(a_d|x|^{2-d}) + O(|x|^{-d-1}),$$

and similarly,

$$\sum_{n=0}^{\infty} 2(\frac{d}{4\pi n})^{d/2}|y|^2 D_{uu} e^{-\frac{d|x|^2}{4n}} = |y|^2 D_{uu}(a_d|x|^{2-d}) + O(|x|^{-d-2}).$$

If $y \not\to 0$, $y \neq -x$,

$$\nabla_y G(x) = \frac{1}{2d} \sum_{|e|=1} \nabla_{y+e} G(x)$$

$$= \frac{1}{2d} \sum_{|e|=1} |y + e|D_u(a_d|x|^{2-d}) + c_y O(|x|^{-d})$$

$$= |y|D_u(a_d|x|^{2-d}) + c_y O(|x|^{-d}),$$

$$\nabla_y^2 G(x) = \frac{1}{2d} \sum_{|e|=1} \nabla_{y+e}^2 G(x)$$

$$= |y|^2 D_{uu}(a_d|x|^{2-d}) + c_y O(|x|^{-d}).$$

Similarly if $x \not\to 0$, we may use (1.8) and (1.9) to show

$$\nabla_y G(x) = \frac{1}{2d} \sum_{|e|=1} \nabla_y G(x + e)$$

$$= |y|D_u(a_d|x|^{2-d}) + c_y O(|x|^{-d-1})$$

$$\nabla_y^2 G(x) = \frac{1}{2d} \sum_{|e|=1} \nabla_y^2 G(x + e)$$

$$= |y|D_{uu}(a_d|x|^{2-d}) + c_y O(|x|^{-d-1}). \quad \square$$

There are two other Green's functions that will be important. Let $\lambda \in [0, 1)$, let T be a geometric random variable independent of S with killing rate $1 - \lambda$ (see Section 1.3), and let $G_\lambda(x, y) = G_\lambda(y - x)$ be the expected number of visits to y starting at x up to the "killing time," i.e.,

$$
\begin{aligned}
G_\lambda(x,y) &= E^x(\sum_{j=0}^{T} I\{S_j = y\}) \\
&= \sum_{j=0}^{\infty} P^x\{S_j = y, T \geq j\} \\
&= \sum_{j=0}^{\infty} \lambda^j p_j(y - x).
\end{aligned}
$$

$G_\lambda(x,y)$ is the generating function of $\{p_j(y - x)\}_{j=0,1,\dots}$. Note that G_λ is finite in all dimensions, and if $d \geq 3$, $\lim_{\lambda \to 1-} G_\lambda(x,y) = G(x,y)$.

Exercise 1.5.6 *Show that if $\lambda > 0$,*

$$
G_\lambda(x,y) = \sum_{j=0}^{\infty} (1 - \lambda)\lambda^j G_j(x,y).
$$

If $A \subset Z^d$, we let $G_A(x,y)$ be the expected number of visits to y starting at x before leaving A. To be precise, let $\tau = \inf\{j \geq 0 : S_j \in \partial A\}$. Then

$$
\begin{aligned}
G_A(x,y) &= E^x[\sum_{j=0}^{\tau-1} I\{S_j = y\}] \\
&= \sum_{j=0}^{\infty} P^x\{S_j = y, \tau > j\}.
\end{aligned}
$$

(Our notation for Green's functions is somewhat ambiguous. In order to distinguish G_n, G_λ, and G_A, one must know whether the subscript is an integer, a real number less than one, or a subset of Z^d. This should not present any problem.) If $A \neq Z^d$, $G_A(x,y)$ will be finite in all dimensions. There is a one-to-one correspondence between random walk paths starting at x and ending at y staying in A and random walk paths starting at y and ending at x staying in A (just traverse the path backwards). Therefore, $P^x\{S_j = y, \tau > j\} = P^y\{S_j = x, \tau > j\}$ and

$$
G_A(x,y) = G_A(y,x).
$$

It is not in general true that $G_A(x,y) = G_A(0, y - x)$; however, if we let $A_x = \{z - x : z \in A\}$, then

$$
G_A(x,y) = G_{A_x}(0, y - x).
$$

Also, if $A \subset B$,

$$
G_A(x,y) \leq G_B(x,y).
$$

Exercise 1.5.7 *Let* $A \subset Z^d$ *be finite,* $x \in A$,

$$\begin{aligned} \tau &= \inf\{j \geq 0 : S_j \in \partial A\} \\ \sigma_x &= \inf\{j \geq 1 : S_j = x\}. \end{aligned}$$

Show that

$$G_A(x,x) = [P^x\{\tau < \sigma_x\}]^{-1}. \tag{1.38}$$

We define the hitting distribution of the boundary of A by

$$H_{\partial A}(x,y) = P^x\{S_\tau = y\}.$$

Proposition 1.5.8 *Let* $A \subset Z^d(d \geq 3)$ *be finite and* $x, z \in A$, *Then*

$$G_A(x,z) = G(z-x) - \sum_{y \in \partial A} H_{\partial A}(x,y)G(z-y).$$

Proof.

$$\begin{aligned} G_A(x,z) &= E^x[\sum_{j=0}^{\tau-1} I\{S_j = z\}] \\ &= E^x[\sum_{j=0}^{\infty} I\{S_j = z\} - \sum_{j=\tau}^{\infty} I\{S_j = z\}] \\ &= G(z-x) - \sum_{y \in \partial A} H_{\partial A}(x,y)G(z-y). \quad \square \end{aligned}$$

We let C_n be the "ball" of radius n about 0, i.e.,

$$C_n = \{z \in Z^d : |z| < n\}.$$

Proposition 1.5.9 *Let* $\eta = \inf\{j \geq 0 : S_j \in \{0\} \cup \partial C_n\}$. *Then if* $x \in C_n$,

$$P^x\{S_\eta = 0\} = \frac{a_d}{G(0)}[|x|^{2-d} - n^{2-d}] + O(|x|^{1-d}),$$

and

$$G_{C_n}(x,0) = a_d(|x|^{2-d} - n^{2-d}) + O(|x|^{1-d}).$$

Proof. Recall that $G(x)$ is harmonic for $x \neq 0$. Therefore, if $S_0 = x$, $M_j = G(S_{j \wedge \eta})$ is a bounded martingale. By the optional sampling theorem,

$$\begin{aligned} G(x) &= E^x(M_\eta) \\ &= G(0)P^x\{S_\eta = 0\} \\ &\quad + E^x(G(S_\eta) \mid S_\eta \in \partial C_n)P^x\{S_\eta \in \partial C_n\}. \end{aligned} \tag{1.39}$$

By Theorem 1.5.4,

$$G(x) = a_d|x|^{2-d} + o(|x|^{1-d}).$$

If $y \in \partial C_n$, then $n \leq |y| < n + 1$, and therefore again by Theorem 1.5.4,

$$E^x(G(S_\eta) \mid S_\eta \in \partial C_n) = a_d n^{2-d} + O(n^{1-d}).$$

If we plug this into (1.39) we get the first result. To get the second, note that

$$G_{C_n}(x, 0) = P^x\{S_\eta = 0\}G_{C_n}(0, 0)$$

and

$$G_{C_n}(0, 0) = G(0) + O(n^{2-d}). \quad \square$$

Proposition 1.5.10 *Suppose* $n < m$ *and* $A = \{z : n < |z| < m\}, \tau = \inf\{j \geq 0 : S_j \in \partial A\}$. *Then for* $x \in A$,

$$P^x\{|S_\tau| \leq n\} = \frac{|x|^{2-d} - m^{2-d} + O(n^{1-d})}{n^{2-d} - m^{2-d}}.$$

Proof. Consider the bounded martingale $M_j = G(S_{j \wedge \tau})$. By the optional sampling theorem,

$$
\begin{aligned}
G(x) &= E^x(M_\tau) \\
&= P^x\{|S_\tau| \leq n\}E^x(G(S_\tau) \mid |S_\tau| \leq n) \\
&\quad + (1 - P^x\{|S_\tau| \leq n\})E^x(G(S_\tau) \mid |S_\tau| \geq m).
\end{aligned}
$$

But by Theorem 1.5.4,

$$
\begin{aligned}
G(x) &= a_d|x|^{2-d} + o(|x|^{1-d}), \\
E^x(G(S_\tau) \mid |S_\tau| \leq n) &= a_d n^{2-d} + O(n^{1-d}), \\
E^x(G(S_\tau) \mid |S_\tau| \geq m\} &= a_d m^{2-d} + O(m^{1-d}).
\end{aligned}
$$

If we solve we get the result. $\square$

Exercise 1.5.11 *Let* $A \subset Z^d$ *be finite,* $F : \partial A \to R$, $g : A \to R$. *Then the unique function* $f : \overline{A} \to R$ *satisfying*

$$(a) \ \Delta f(x) = -g(x), \ x \in A,$$

$$(b) \ f(x) = F(x), \ x \in \partial A,$$

is

$$f(x) = \sum_{y \in \partial A} H_{\partial A}(x, y)F(y) + \sum_{z \in A} g(z)G_A(x, z).$$

1.6   Recurrent Case

The Green's function $G(x, y)$ is infinite if $d \leq 2$. However, there is another useful quantity called the *potential kernel* defined by

$$
\begin{aligned}
a(x) & = \lim_{n \to \infty} [G_n(0) - G_n(x)] \\
& = \lim_{n \to \infty} \sum_{j=0}^{n} (p_j(0) - p_j(x)).
\end{aligned} \tag{1.40}
$$

Theorem 1.6.1 *If $d \leq 2$, the limit in (1.40) is finite, i.e., $a(x)$ is well-defined.*

Proof. For each n, if $x \neq 0, x \leftrightarrow 0$,

$$
\sum_{j=0}^{n} (p_j(0) - p_j(x)) = 1 + \sum_{1 \leq j \leq n/2} (\bar{p}_{2j}(0) - \bar{p}_{2j}(x)) + \sum_{j=1}^{n} E(j, 0) - \sum_{j=1}^{n} E(j, x).
$$

Since $\bar{p}_{2j}(0) - \bar{p}_{2j}(x) = O(j^{-d/2})(1 - \exp\{-\frac{d|x|^2}{4j}\}) = |x|^2 O(j^{-(d+2)/2})$,

$$
\sum_{j=1}^{\infty} |\bar{p}_{2j}(0) - \bar{p}_{2j}(x)| < \infty.
$$

By (1.10),

$$
\sum_{j=1}^{\infty} |E(j, 0)| < \infty, \quad \sum_{j=1}^{\infty} |E(j, x)| < \infty.
$$

Therefore the limit exists and

$$
a(x) = \sum_{j=0}^{\infty} (\bar{p}_{2j}(0) - \bar{p}_{2j}(x)) + \sum_{j=1}^{\infty} E(j, 0) - \sum_{j=1}^{\infty} E(j, x). \tag{1.41}
$$

If $x \not\leftrightarrow 0$, then

$$
G_n(x) = \frac{1}{2d} \sum_{|e|=1} G_{n-1}(x + e),
$$

and therefore,

$$
\begin{aligned}
\lim_{n \to \infty} (G_n(0) - G_n(x)) & = \lim_{n \to \infty} \frac{1}{2d} \sum_{|e|=1} [G_{n-1}(0) - G_{n-1}(x + e)] + p_n(0) \\
& = \frac{1}{2d} \sum_{|e|=1} a(x + e). \quad \square
\end{aligned}
$$

It is easy to see that $a(x)$ satisfies

$$a(x) = a(-x),$$

$$\Delta a(x) = \delta(x).$$

The next theorem gives the asymptotic behavior of $a(x)$ as $|x| \to \infty$ for $d = 2$. (In Theorem 1.6.4 we will compute $a(x)$ exactly for $d = 1$.) Note that the function $f(x) = \ln |x|$ is a harmonic function on $R^2 \backslash \{0\}$. Therefore by (1.9), if f is considered as a function on Z^2,

$$\Delta f(x) = O(|x|^{-3}). \tag{1.42}$$

Theorem 1.6.2 *If $d = 2$, there exists a constant k such that if $\alpha < 2$*

$$\lim_{|x| \to \infty} |x|^\alpha [a(x) - \frac{2}{\pi} \ln |x| - k] = 0.$$

Proof. By (1.41), if $x \leftrightarrow 0, x \neq 0$,

$$a(x) = \sum_{j=1}^{\infty} (\bar{p}_{2j}(0) - \bar{p}_{2j}(x)) + 1 + \sum_{j=1}^{\infty} E(j, 0) - \sum_{j=1}^{\infty} E(j, x).$$

We split the first term into two parts and estimate separately.

$$\sum_{j=1}^{\infty} (\bar{p}_{2j}(0) - \bar{p}_{2j}(x)) = \sum_{1 \leq j \leq |x|^2} \frac{1}{\pi j} (1 - e^{-\frac{|x|^2}{2j}}) + \sum_{j > |x|^2} \frac{1}{\pi j} (1 - e^{-\frac{|x|^2}{2j}}).$$

$$\sum_{1 \leq j \leq |x|^2} \frac{1}{\pi j} (1 - e^{-\frac{|x|^2}{2j}}) = [\sum_{1 \leq j \leq |x|^2} \frac{1}{\pi j}] - \int_0^{|x|^2} \frac{1}{\pi t} e^{-\frac{|x|^2}{2t}} dt + o(|x|^{-2}).$$

The first term on the right hand side equals $\frac{1}{\pi} \ln |x|^2 + \gamma + O(|x|^{-2})$, where γ is Euler's constant. The substitution $u = (2t)^{-1} |x|^2$ in the integral then gives

$$\sum_{1 \leq j \leq |x|^2} \frac{1}{\pi j} (1 - e^{-\frac{|x|^2}{2j}}) = \frac{2}{\pi} \ln |x| + \gamma - \int_{\frac{1}{2}}^{\infty} \frac{1}{\pi u} e^{-u} du + O(|x|^{-2}).$$

For the second part,

$$\sum_{|x|^2 < j} \frac{1}{\pi j} (1 - e^{-\frac{|x|^2}{2j}}) = O(|x|^{-2}) + \int_{|x|^2}^{\infty} \frac{1}{\pi t} (1 - e^{-\frac{|x|^2}{2t}}) dt.$$

$$= O(|x|^{-2}) + \int_0^{\frac{1}{2}} \frac{1}{\pi u} (1 - e^{-u}) du.$$

By (1.30), if $\alpha < 2$,

$$\lim_{|x|\to\infty} |x|^\alpha [\sum_{j=0}^{\infty} |E(j,x)|] = 0.$$

We have therefore proved the theorem for $x \leftrightarrow 0$ with

$$k = \gamma - \int_{\frac{1}{2}}^{\infty} \frac{1}{\pi u} e^{-u} du + \int_0^{\frac{1}{2}} \frac{1}{\pi u}(1 - e^{-u}) du + 1 + \sum_{j=1}^{\infty} E(j,0).$$

If $x \not\leftrightarrow 0$, we may use $\Delta a(x) = 0$ and (1.42) to obtain

$$
\begin{aligned}
a(x) &= \frac{1}{2d} \sum_{|e|=1} a(x+e) \\
&= \frac{2}{\pi} \ln |x| + \frac{2}{\pi}\Delta(\ln|x|) + k + o(|x|^{-\alpha}) \\
&= \frac{2}{\pi} \ln |x| + k + o(|x|^{-\alpha}). \quad \square
\end{aligned}
$$

The above theorem can be improved to show

$$a(x) = \frac{2}{\pi} \ln |x| + k + O(|x|^{-2}),$$

(see [67]), but we will not need this stronger result. Also the value of k can be calculated [65, p. 124],

$$k = \frac{2\gamma}{\pi} + \frac{3}{\pi} \ln 2,$$

although we will not need to know this value. The potential kernel can often be used in place of the Green's function, e.g., this next proposition is the analogue to Proposition 1.5.8.

Proposition 1.6.3 If $A \subset Z^d$ $(d \leq 2)$ is finite, then if $x, z \in A$,

$$G_A(x,z) = [\sum_{y \in \partial A} H_{\partial A}(x,y)a(y-z)] - a(z-x).$$

Proof. We could give a proof similar to that of Proposition 1.5.8. For variety, however, we note that $h(x) = a(x-z)$ satisfies $\Delta h(x) = \delta(z-x)$ and hence by (1.23),

$$h(x) = E^x[h(S_\tau)] - E^x[\sum_{j=0}^{\tau-1} I\{S_j = z\}]. \quad \square$$

Theorem 1.6.4 *If $d = 1$, $a(x) = |x|$.*

Proof. Let $A = \{z \in Z : |z| < |x|\}$, $\partial A = \{-x, x\}$. Since $a(x) = a(-x)$, Proposition 1.6.3 implies that $G_A(0,0) = a(x)$. But by (1.38) and (1.20),

$$[G_A(0,0)]^{-1} = P^0\{\tau < \sigma_0\} = |x|^{-1}. \quad \Box$$

Theorem 1.6.5 *If $y \in Z^2, y = |y|u$, then*

$$(a)\ |\nabla_y a(x) - |y|D_u(\frac{2}{\pi}\ln|x|)| \quad = \quad O(|x|^{-2}),$$

$$(b)\ |\nabla_y^2 a(x) - |y|^2 D_{uu}(\frac{2}{\pi}\ln|x|)| \quad = \quad O(|x|^{-3}).$$

Proof. Identical to the proof of Theorem 1.5.5. $\Box$

Theorem 1.6.6 *Let $C_n = \{z \in Z^2 : |z| < n\}$. Then*

$$G_{C_n}(0,0) = \frac{2}{\pi}\ln n + k + O(n^{-1}).$$

Proof. By Proposition 1.6.3,

$$G_{C_n}(0,0) = \sum_{y \in \partial C_n} H_{\partial C_n}(0,y)a(y).$$

Since $n \le |y| < n+1$ for $y \in \partial C_n$, Theorem 1.6.2 gives the result. $\Box$

Proposition 1.6.7 *Let $x \in C_n$ and $\eta = \inf\{j \ge 1 : S_j \in \{0\} \cup \partial C_n\}$. Then for every $\alpha < 2$,*

$$P^x\{S_\eta = 0\} = (\ln n)^{-1}[\ln n - \ln|x| + o(|x|^{-\alpha}) + O((\ln n)^{-1})].$$

$$G_{C_n}(x,0) = \frac{2}{\pi}[\ln n - \ln|x|] + o(|x|^{-\alpha}) + O(n^{-1}).$$

Proof. Assume $S_0 = x$. Then $M_j = a(S_{j \wedge \eta})$ is a bounded martingale and by the optional sampling theorem,

$$a(x) \quad = \quad E^x(M_\eta)$$
$$= \quad (1 - P^x\{S_\eta = 0\})E^x(a(S_\eta) \mid |S_\eta| \ge n).$$

But by Theorem 1.6.2

$$a(x) = \frac{2}{\pi}\ln|x| + k + o(|x|^{-1}),$$

and hence,

$$E^x(a(S_\eta) \mid |S_\eta| \geq n) = \frac{2}{\pi} \ln n + k + O(n^{-1}).$$

If we solve for $P^x\{S_\eta = 0\}$, we get

$$P^x\{S_\eta = 0\} = \frac{\frac{2}{\pi}(\ln n - \ln|x|) + o(|x|^{-\alpha})}{\frac{2}{\pi}\ln n + k + O(n^{-1})}.$$

This gives the first equation. The second equation then follows from

$$G_{C_n}(x,0) = P^x\{S_\eta = 0\}G_{C_n}(0,0)$$

and Theorem 1.6.6. □

Exercise 1.6.8 *Suppose $n < m$, $A = \{x \in Z^2 : n < |x| < m\}$, $\tau = \inf\{j \geq 1 : S_j \in \partial A\}$. Then for $x \in A$,*

$$P^x\{|S_\tau| \leq n\} = \frac{\ln m - \ln|x| + O(n^{-1})}{\ln m - \ln n}.$$

1.7 Difference Estimates and Harnack Inequality

Let $B = \{x \in R^d : |x| < 1\}$, $\overline{B} = \{x \in R^d : |x| \leq 1\}$, and $f : \overline{B} \to R$ a function which is harmonic , i.e., $\sum_{i=1}^d D_{ii}f(x) = 0$ for $x \in B$. Then the Poisson integral formula [28] states that

$$f(x) = \int_{\partial B} f(s)\rho(s,x)ds, \tag{1.43}$$

where $\rho(s,x)$ is the Poisson kernel

$$\rho(s,x) = c_d \frac{1 - |x|^2}{|s - x|^d}. \tag{1.44}$$

From this formula we can derive some immediate estimates. For example, if we differentiate f at the origin in the direction y:

$$|D_y f(0)| \leq \|f\|_\infty \int_{\partial B} |D_y \rho(s,0)|ds \leq c\|f\|_\infty \sup_{s \in \partial B} |D_y \rho(s,0)|.$$

In particular, the derivative is bounded by the supremum of f times a constant independent of f. Another corollary of (1.43) is the Harnack inequality. Suppose $f \geq 0$ on D. If $|x|, |y| \leq r < 1$, then for each $s \in \partial D$,

$$\rho(s,x) \leq c_d(1 - r^2)^{-1}[\frac{1 + r}{1 - r}]^d \rho(s,y),$$

and hence

$$f(x) \leq c_d(1 - r^2)^{-1}[\frac{1+r}{1-r}]^d f(y).$$

Note that the constant does not depend on f. For discrete harmonic functions we do not have as explicit a formula as (1.43) and (1.44); however, it will be useful to have bounds on differences and to have a Harnack inequality. We derive such results in this section. We start by stating the results we will prove. As before, let

$$C_n = \{z \in Z^d : |z| < n\}.$$

Theorem 1.7.1 *For each* $u \in Z^d$, *there exists a* $c_u < \infty$ *such that if* $f : \overline{C}_n \to R$ *is harmonic on* C_n,

$$\begin{aligned} (a)\ |\nabla_u f(0)| &\leq c_u \|f\|_\infty O(n^{-1}), \\ (b)\ |\nabla_u^2 f(0)| &\leq c_u \|f\|_\infty O(n^{-2}). \end{aligned}$$

Theorem 1.7.2 (Harnack inequality) *For each* $r < 1$, *there exists a* $c_r < \infty$ *such that if* $f : \overline{C}_n \to [0, \infty)$ *is harmonic on* C_n,

$$f(x_1) \leq c_r f(x_2), \quad |x_1|, |x_2| \leq rn.$$

If $x \in A$, $V \subset A$, we say V separates x from ∂A if every path from x to ∂A must enter some point of V, i.e., if $P^x\{\tau_V < \tau\} = 1$ where $\tau_V = \inf\{j \geq 0 : S_j \in V\}, \tau = \inf\{j \geq 0 : S_j \in \partial A\}$. Suppose $B \subset A$, $\partial B \subset A$. Then it is easy to check that ∂B separates any point of B from ∂A.

Lemma 1.7.3 *Suppose* $A \subset Z^d$ *is finite and* $V \subset A$ *separates* x *from* ∂A. *Let* $\tau = \inf\{j \geq 0 : S_j \notin A\}$ *and*

$$H(x, y) = H_{\partial A}(x, y) = P^x\{S_\tau = y\}.$$

Let $\overline{\tau} = \inf\{j \geq 1 : S_j \in \partial A \cup V\}$. *Then*

$$H(x, y) = \sum_{z \in V} P^y\{S_{\overline{\tau}} = z\} G_A(z, x) \tag{1.45}$$

Proof. Let

$$\eta = \inf\{j \geq 1 : S_j \in \{x\} \cup \partial A\}.$$

By considering paths in reverse direction we can see that if $y \in \partial A$,

$$P^y\{S_\eta = x\} = P^x\{S_\eta = y\}.$$

Also by the strong Markov property,

$$P^x\{S_\tau = y \mid S_\eta = x\} = P^x\{S_\tau = y\}.$$

Hence $P^x\{S_\eta = y \mid S_\eta \neq x\} = P^x\{S_\tau = y\}$ and

$$P^x\{S_\eta = y\} = P^x\{S_\eta \neq x\}H(x,y).$$

Therefore, using (1.38),

$$
\begin{aligned}
H(x,y) &= [P^x\{S_\eta \neq x\}]^{-1}P^x\{S_\eta = y\} \\
&= G_A(x,x)P^y\{S_\eta = x\} \\
&= G_A(x,x)\sum_{z \in V} P^y\{S_{\overline{\tau}} = z\}P^z\{S_\eta = x\}.
\end{aligned}
$$

By the strong Markov property, if $z \in A$, $G_A(z,x) = P^z\{S_\eta = x\}G_A(x,x)$. Therefore,

$$H(x,y) = \sum_{z \in V} P^y\{S_{\overline{\tau}} = z\}G_A(z,x). \quad \square$$

Lemma 1.7.4 If $y \in \partial C_n$,

$$H(0,y) \asymp n^{1-d}.$$

Proof. It suffices to prove the lemma for $d \geq 2$ and n sufficiently large. Let $\eta = \inf\{j \geq 1 : S_j \in \{0\} \cup \partial C_n\}$. Then as in the previous proof,

$$H(0,y) = G_{C_n}(0,0)P\{S_\eta = y\} = G_{C_n}(0,0)P^y\{S_\eta = 0\}.$$

Let $\rho = \inf\{j \geq 1 : |S_j| < n-2 \text{ or } |S_j| \geq n\}$. Then by the strong Markov property,

$$\underline{a}P^y\{|S_\rho| < n-2\} \leq P^y\{S_\eta = 0\} \leq \overline{a}P^y\{|S_\rho| < n-2\},$$

where $\underline{a}$ ($\overline{a}$) is the infimum (supremum) of $P^z\{S_\eta = 0\}$ over all $z \in \partial C_{n-3}$. By Proposition 1.5.9 or Proposition 1.6.7,

$$\underline{a} \asymp n^{1-d}, \qquad \overline{a} \asymp n^{1-d}, \qquad d \geq 3,$$

$$\underline{a} \asymp (n\ln n)^{-1}, \qquad \overline{a} \asymp (n\ln n)^{-1}, \qquad d = 2.$$

Therefore,

$$P^y\{S_\eta = 0\} \asymp n^{1-d}P^y\{|S_\rho| < n-2\}, \quad d \geq 3,$$

$$P^y\{S_\eta = 0\} \asymp (n\ln n)^{-1}P^y\{|S_\rho| < n-2\}, \quad d = 2.$$

Since $G_{C_n}(0,0) \asymp 1$ if $d \geq 3$ and $G_{C_n}(0,0) \asymp \ln n$ if $d = 2$ (Theorem 1.6.6), it suffices to prove that $P^y\{|S_\rho| < n-2\} \geq c > 0$. For $y \in \partial C_n$, choose $|e| = 1$ which minimizes $|y - e|$, i.e., which maximizes $y \cdot e$. We know $|y - e| < n$ and by simple geometry $y \cdot e \geq |y| d^{-1}$. Then for any k,

$$|y - ke|^2 = |y|^2 + k^2 - 2k(y \cdot e) \leq |y|^2 + k^2 - 2k|y|d^{-1}.$$

Let $k = 4d$. Then for $|y|$ sufficiently large,

$$|y - ke|^2 < (|y| - 3)^2 \leq (n - 2)^2,$$

and hence $y - ke \in C_{n-2}$. Since $y - e, y - 2e, \ldots, y - ke \in C_n$,

$$P^y\{|S_\rho| < n - 2\} \geq (2d)^{-k} = (2d)^{-4d}. \quad \square$$

Proof of Theorem 1.7.1. Let $V_n = C_{n/2}$. Then if $n \geq 3$, $\partial V_n \subset C_n$ and hence ∂V_n separates 0 from ∂C_n. For $z \in \partial V_n$,

$$\frac{n}{2} \leq |z| < \frac{n}{2} + 1.$$

If $d = 2$, Proposition 1.6.3 gives

$$G_A(z, x) = \sum_{y \in \partial A} H(z, y)(a(y - x) - a(z - x)),$$

while if $d \geq 3$, Proposition 1.5.8 gives

$$G_A(z, x) = \sum_{y \in \partial A} H(z, y)(G(z - x) - G(y - x)),$$

where $A = C_n$. By Theorem 1.5.5 or Theorem 1.6.5, if $g(x) = G_A(z, x)$, then

$$|\nabla_u g(0)| \leq c_u n^{1-d},$$
$$|\nabla_u^2 g(0)| \leq c_u n^{-d}.$$

For $z \in \partial V_n$, $G_A(z, 0) \geq cn^{2-d}$ (Proposition 1.5.9, Proposition 1.6.7), therefore

$$|\nabla_u g(0)| \leq c_u O(n^{-1}) g(0)$$
$$|\nabla_u^2 g(0)| \leq c_u O(n^{-2}) g(0).$$

By (1.45) and Lemma 1.7.4,

$$|\nabla_u H(0, y)| \leq c_u O(n^{-1}) H(0, y) \leq cn^{-d},$$
$$|\nabla_u^2 H(0, y)| \leq c_u O(n^{-2}) H(0, y) \leq cn^{-d-1},$$

where on the left hand side the differences are taken with respect to the first component of $H(x, y)$. By (1.22),

$$f(x) = \sum_{y \in \partial C_n} f(y) H(x, y),$$

and hence

$$
\begin{aligned}
|\nabla_u f(0)| &\leq \|f\|_\infty \sum_{y \in \partial C_n} |\nabla_u H(0, y)| \\
&\leq c_u \|f\|_\infty O(n^{-1}), \\
|\nabla_u^2 f(0)| &\leq \|f\|_\infty \sum_{y \in \partial C_n} |\nabla_u^2 H(0, y)| \\
&\leq c_u \|f\|_\infty O(n^{-2}). \quad \square
\end{aligned}
$$

Exercise 1.7.5 *Show that if $f : \overline{C}_n \to R$ is harmonic, then for every u,*

$$|\nabla_u f(0)| \leq |u| \|f\|_\infty O(n^{-1}).$$

(Hint: It suffices to consider $|u| \leq \frac{n}{2}$. Use the estimate from Theorem 1.7.1(a) with $|u| = 1$ and iterate.)

Proof of Theorem 1.7.2. We will first prove the result for $r = \frac{1}{16}$. Assume $z \in \partial C_{n/4}, x \in \overline{C}_{n/16}$. Then

$$G_{C_{2n/3}}(0, z - x) \leq G_{C_n}(z, x) \leq G_{C_{3n/2}}(0, z - x).$$

Also, $\frac{3n}{16} - 1 \leq |z - x| \leq \frac{5n}{16} + 1$. Therefore, by Proposition 1.5.10 or Proposition 1.6.7, there exists a constant c such that for every $z_1, z_2 \in \partial C_{n/4}, x_1, x_2 \in \overline{C}_{n/16}$,

$$G_{C_n}(z_1, x_1) \leq c\, G_{C_n}(z_2, x_2).$$

If $y \in \partial C_n$, Lemma 1.7.3 gives

$$P^x\{S_\tau = y\} = \sum_{z \in \partial C_{n/4}} P^y\{S_{\overline{\tau}} = z\} G_{C_n}(z, x).$$

Therefore, if $x_1, x_2 \in \overline{C}_{n/16}$, $H(x_1, y) \leq cH(x_2, y)$ and by (1.22), $f(x_1) \leq cf(x_2)$.

Now let $r < 1$ and $|x_1|, |x_2| \leq rn$. If $|x_1 - x_2| \leq \frac{1}{16}(1 - r)n$, then we can apply the above result to $C_{(1-r)n}(x_1) = \{z : z - x \in C_{(1-r)n}\}$ to get

$f(x_1) \leq cf(x_2)$. By induction, if $|x_1|, \ldots, |x_{k+1}| \leq rn$ and $|x_{j+1} - x_j| \leq \frac{1}{16}(1 - r)n, j = 1, 2, \ldots, k$, then

$$f(x_1) \leq c^k f(x_{k+1}).$$

If k_r is any integer greater than $32(1 - r)^{-1}$, then for n sufficiently large, if $|x_1|, |x_2| \leq rn$, there exist $z_1 = x_1, z_2, \ldots, z_{k_r+1} = x_2$ with $|z_j| \leq rn$ and $|z_{j+1} - z_j| \leq \frac{1}{16}(1 - r)n$. Therefore,

$$f(x_1) \leq c^{k_r} f(x_2). \quad \square$$

The second part of the proof of Theorem 1.7.2 can be applied to more general sets than $C_{(1-r)n}$. If U is a compact subset of R^d contained in an open set V, then U can be covered by a finite number of open balls with centers in U and radius at most $\frac{1}{2}\text{dist}(U, \partial V)$. Using this idea, we can prove the following result, which we will refer to as the Harnack principle. We leave the proof as an exercise.

Theorem 1.7.6 (Harnack principle) *Let U be a compact subset of R^d contained in a connected open set V. Then there exists a $c = c(U, V) < \infty$ such that if $A_n = nU \cap Z^d$, $B_n = nV \cap Z^d$, and $f : \overline{B}_n \to [0, \infty)$ is harmonic in B_n, then*

$$f(x) \leq cf(y), \ x, y \in A_n.$$

Chapter 2

Harmonic Measure

2.1 Definition

The hitting probability of a set $A \in Z^d$ is the function $H_A : Z^d \times A \to [0,1]$ defined by

$$H_A(x,y) = P^x\{S_\tau = y\},$$

where

$$\tau = \tau_A = \inf\{j \geq 1 : S_j \in A\}.$$

This differs from the definition of τ_A in Chapter 1 where we took the infimum over $j \geq 0$, but it will be more useful in this chapter to have τ_A defined as above. Note that $\tau_{A \cup B} = \tau_A \wedge \tau_B$, and by "reversing paths" we can see that if $x, y \in A$,

$$H_A(x,y) = H_A(y,x). \tag{2.1}$$

For fixed $y \in A$, $H_A(\cdot, y)$ is a harmonic function in $\overline{A}^c$. For fixed $x \in Z^d$, $H_A(x, \cdot)$ is a positive measure on A with total mass $P^x\{\tau < \infty\}$. We may define a probability measure on A by conditioning that the random walk hits A,

$$\overline{H}_A(x,y) = P^x\{S_\tau = y \mid \tau < \infty\}.$$

For $d \leq 2$, $H_A(x,y) = \overline{H}_A(x,y)$ by recurrence.

If A is a finite set, we define the harmonic measure of A, $H_A(\cdot)$, to be the hitting probability from infinity, i.e.,

$$H_A(y) = \lim_{|x| \to \infty} \overline{H}_A(x,y). \tag{2.2}$$

In this section we will show that H_A is well defined by showing the limit exists, and in the process we will relate H_A to probabilities of "escaping"

47

the set A. We start by developing some properties of $H_A(x, y)$. Let $C_n = \{z \in Z^d : |z| < n\}$ and assume $A \subset C_n$. If $m \geq n$ and $x \notin \overline{C}_m$, then ∂C_m separates x from A, i.e., every path from x to A must go through some point of ∂C_m.

Lemma 2.1.1 *If $A \subset B$ and $x \in B^c$, then*

$$H_A(x, y) = \sum_{z \in \partial B} G_{A^c}(x, z) H_{A \cup \partial B}(z, y),$$

$$\overline{H}_A(x, y) = \frac{\sum_{z \in \partial B} G_{A^c}(x, z) H_{A \cup \partial B}(z, y)}{\sum_{z \in \partial B} G_{A^c}(x, z) P^z\{\tau_A < \tau_{\partial B}\}}. \qquad (2.3)$$

Proof. Consider the random time

$$\sigma = \sup\{j < \tau_A : S_j \in \partial B\}.$$

Note that σ is not a stopping time. However, since $\sigma < \tau_A$,

$$
\begin{aligned}
P^x\{S_{\tau_A} = y\} &= \sum_{k=1}^{\infty} P^x\{\tau_A = k, S_k = y\} \\
&= \sum_{k=1}^{\infty} \sum_{j=0}^{k-1} \sum_{z \in \partial B} P^x\{\tau_A = k, S_k = y, \sigma = j, S_j = z\} \\
&= \sum_{z \in \partial B} \sum_{j=0}^{\infty} \sum_{k=j+1}^{\infty} P^x\{S_j = z; S_k = y; S_i \notin A, \\
&\qquad\qquad 0 \leq i \leq j; S_i \notin A \cup \partial B, j + 1 \leq i < k\} \\
&= \sum_{z \in \partial B} \sum_{j=0}^{\infty} P^x\{S_j = z; S_i \notin A, 0 \leq i \leq j\} \\
&\qquad\qquad \sum_{k=1}^{\infty} P^z\{S_k = y; S_i \notin A \cup \partial B, 1 \leq i < k\} \\
&= \sum_{z \in \partial B} G_{A^c}(x, z) H_{A \cup \partial B}(z, y).
\end{aligned}
$$

Similarly, by summing over $y \in A$,

$$P^x\{\tau_A < \infty\} = \sum_{z \in \partial B} G_{A^c}(x, z) P^z\{\tau_A < \tau_{\partial B}\}.$$

which gives (2.3). $\square$

An immediate consequence of (2.3) is that for $x \in B^c$,

$$\inf_{z \in \partial B} \frac{H_{A \cup \partial B}(z, y)}{P^z\{\tau_A < \tau_{\partial B}\}} \leq \overline{H}_A(x, y) \leq \sup_{z \in \partial B} \frac{H_{A \cup \partial B}(z, y)}{P^z\{\tau_A < \tau_{\partial B}\}},$$

or by (2.1),

$$\inf_{z \in \partial B} \frac{H_{A \cup \partial B}(y, z)}{\sum_{y \in A} P^y\{S(\tau_A \wedge \tau_{\partial B}) = z\}} \leq \overline{H}_A(x, y) \tag{2.4}$$

$$\leq \sup_{z \in \partial B} \frac{H_{A \cup \partial B}(y, z)}{\sum_{y \in A} P^y\{S(\tau_A \wedge \tau_{\partial B}) = z\}}.$$

We will use this inequality with $B = C_m, m > n$.

Lemma 2.1.2 *Assume $A \subset C_n$, $B = C_m$, and $m > 4n$. Then for $y \in A$, $z \in \partial B$,*

$$P^y\{S(\tau_A \wedge \tau_{\partial B}) = z \mid \tau_A > \tau_{\partial B}\} = \begin{cases} H_{\partial B}(0, z)(1 + O(\frac{n}{m})), & d \geq 3, \\ H_{\partial B}(0, z)(1 + O(\frac{n}{m} \ln \frac{m}{n})), & d = 2. \end{cases}$$

Proof. Since $m > 4n$, $\partial C_{2n} \subset B$. If $d \geq 3$, by Proposition 1.5.10, if $w \in \partial C_{2n}$,

$$P^w\{\tau_A > \tau_{\partial B}\} \geq P^w\{\tau_A = \infty\} \geq c. \tag{2.5}$$

If $d = 2$, Exercise 1.6.8 gives

$$P^w\{\tau_A \geq \tau_{\partial B}\} \geq cO((\ln \frac{m}{n})^{-1}). \tag{2.6}$$

If we consider $f(x) = P^x\{S_{\tau_{\partial B}} = z\}$ as a harmonic function on $C_{3m/4}$, then by Exercise 1.7.5, for $w \in \overline{C}_{2n}$,

$$|f(w) - f(0)| \leq O(\frac{n}{m}) \sup_{x \in \overline{C}_{3m/4}} f(x).$$

But by the Harnack inequality (Theorem 1.7.2) applied to f on $\overline{C}_m$,

$$\sup_{x \in \overline{C}_{3m/4}} f(x) \leq c\, f(0).$$

Therefore, if $w \in \overline{C}_{2n}$,

$$P^w\{S(\tau_{\partial B}) = z\} = H_{\partial B}(0, z)(1 + O(\frac{n}{m})). \tag{2.7}$$

Suppose $\tau_A < \tau_{\partial B}$ and let $\eta = \inf\{j \geq \tau_A : S_j \in \partial C_{2n}\}$. Then by (2.7),

$$
\begin{aligned}
P^w\{S(\tau_{\partial B}) = z \mid \tau_A < \tau_{\partial B}\} \\
= \sum_{x \in \partial C_{2n}} P^w\{S_\eta = x \mid \tau_A < \tau_{\partial B}\} \, P^x\{S(\tau_{\partial B}) = z\} \\
= H_{\partial B}(0, z)(1 + O(\frac{n}{m})).
\end{aligned}
\tag{2.8}
$$

Therefore by (2.5) - (2.8), if $w \in C_{2n}$,

$$
P^w\{S(\tau_{\partial B}) = z \mid \tau_A > \tau_{\partial B}\} = \left\{ \begin{array}{ll} H_{\partial B}(0, z)(1 + O(\frac{n}{m})) & d \geq 3, \\ H_{\partial B}(0, z)(1 + O(\frac{n}{m} \ln \frac{m}{n})) & d = 2, \end{array} \right.
$$

and therefore $P^w\{S(\tau_A \wedge \tau_{\partial B}) = z\}$ equals

$$
H_{\partial B}(0, z)P^w\{\tau_A > \tau_{\partial B}\}(1 + O(\frac{n}{m})) \quad d \geq 3, \tag{2.9}
$$

$$
H_{\partial B}(0, z)P^w\{\tau_A > \tau_{\partial B}\}(1 + O(\frac{n}{m} \ln \frac{m}{n})) \quad d = 2. \tag{2.10}
$$

Since ∂C_{2n} separates y from ∂B, if $d \geq 3$,

$$
\begin{aligned}
P^y\{S(\tau_A \wedge \tau_{\partial B}) = z\} \\
= \sum_{w \in \partial C_{2n}} P^y\{S(\tau_A \wedge \tau_{\partial C_{2n}}) = w\}P^w\{S(\tau_A \wedge \tau_{\partial B}) = z\} \\
= H_{\partial B}(0, z)(1 + O(\frac{n}{m})) \\
\qquad \sum_{w \in \partial C_{2n}} P^y\{S(\tau_A \wedge \tau_{\partial C_{2n}}) = w\}P^w\{\tau_A \geq \tau_{\partial B}\} \\
= H_{\partial B}(0, z)P^y\{\tau_A > \tau_{\partial B}\}(1 + O(\frac{n}{m})).
\end{aligned}
$$

Similarly, if $d = 2$,

$$
P^y\{S(\tau_A \wedge \tau_{\partial B}) = z\} = H_{\partial B}(0, z)P^y\{\tau_A > \tau_{\partial B}\}(1 + O(\frac{n}{m} \ln \frac{m}{n})),
$$

which gives the lemma. □

If $d \geq 3$, and we sum over $y \in A$,

$$
\sum_{y \in A} P^y\{S(\tau_A \wedge \tau_{\partial B}) = z\} = [\sum_{y \in A} P^y\{\tau_A > \tau_{\partial B}\}]H_{\partial B}(0, z)(1 + O(\frac{n}{m})).
$$

Also by Lemma 2.1.2,

$$
H_{A \cup \partial B}(y, z) = P^y\{\tau_A > \tau_{\partial B}\}H_{\partial B}(0, z)(1 + O(\frac{n}{m})).
$$

Therefore by (2.4), if $x \in C_m^c$,

$$\overline{H}_A(x,y) = \frac{P^y\{\tau_A > \tau_{\partial B}\}}{\sum_{\tilde{y} \in A} P^{\tilde{y}}\{\tau_A > \tau_{\partial B}\}}(1 + O(\frac{n}{m})).$$

Similarly for $d = 2$,

$$\overline{H}_A(x,y) = \frac{P^y\{\tau_A > \tau_{\partial B}\}}{\sum_{\tilde{y} \in A} P^{\tilde{y}}\{\tau_A > \tau_{\partial B}\}}(1 + O(\frac{n}{m} \ln \frac{m}{n})).$$

We have proved the following theorem.

Theorem 2.1.3 *Assume $A \subset C_n$. For each $m > n$ define the probability measure*

$$H_A^m(y) = \frac{P^y\{\tau_A > \tau_{\partial C_m}\}}{\sum_{\tilde{y} \in A} P^{\tilde{y}}\{\tau_A > \tau_{\partial C_m}\}}.$$

Then for all $x \in C_m^c, y \in A, m \geq 4n$,

$$\overline{H}_A(x,y) = \begin{cases} H_A^m(y)(1 + O(\frac{n}{m})), & d \geq 3, \\ H_A^m(y)(1 + O(\frac{n}{m} \ln \frac{m}{n})), & d = 2. \end{cases}$$

In particular, the limit in (2.2) exists and

$$\lim_{|x| \to \infty} \overline{H}_A(x,y) = \lim_{m \to \infty} H_A^m(y) = H_A(y).$$

Exercise 2.1.4 *There exist constants c_1, c_2 such that if $A \subset C_n$, $y \in A, m \geq 2n$,*

$$c_1 H_A^m(y) \leq H_A(y) \leq c_2 H_A^m(y).$$

(Hint: use the Harnack principle.)

Exercise 2.1.5 *If $A \subset B$ are finite subsets of Z^d and $x \in A$,*

$$H_A(x) \geq H_B(x). \tag{2.11}$$

2.2 Capacity, Transient Case

For this section we assume $d \geq 3$. If $A \subset Z^d$, we define the escape probability $\mathrm{Es}_A : A \to [0,1]$ by

$$\mathrm{Es}_A(x) = P^x\{\tau_A = \infty\} = \lim_{m \to \infty} P^x\{\tau_A > \xi_m\},$$

where

$$\xi_m = \tau_{\partial C_m}.$$

The *capacity* of a finite set A is given by

$$\text{cap}(A) = \sum_{x \in A} \text{Es}_A(x) = \lim_{m \to \infty} \sum_{x \in A} P^x\{\tau_A > \xi_m\}. \qquad (2.12)$$

It is easy to see that if A is finite, then there exists some $x \in A$ with $\text{Es}_A(x) > 0$, and hence $\text{cap}(A) > 0$. By Theorem 2.1.3,

$$H_A(x) = \frac{\text{Es}_A(x)}{\text{cap}(A)}. \qquad (2.13)$$

Proposition 2.2.1 *Suppose A and B are finite subsets of Z^d ($d \geq 3$).*
(a) If $A \subset B$, then $\text{cap}(A) \leq \text{cap}(B)$.
(b) For any A, B,

$$\text{cap}(A) + \text{cap}(B) \geq \text{cap}(A \cup B) + \text{cap}(A \cap B).$$

Proof: By (2.1) and (2.12),

$$
\begin{aligned}
\text{cap}(A) &= \lim_{m \to \infty} \sum_{y \in A} P^y\{\tau_A > \xi_m\} \\
&= \lim_{m \to \infty} \sum_{y \in A} \sum_{x \in \partial C_m} P^y\{S(\tau_A \wedge \xi_m) = x\} \\
&= \lim_{m \to \infty} \sum_{x \in \partial C_m} \sum_{y \in A} P^x\{S(\tau_A \wedge \xi_m) = y\} \\
&= \lim_{m \to \infty} \sum_{x \in \partial C_m} P^x\{\tau_A < \xi_m\}. \qquad (2.14)
\end{aligned}
$$

Therefore, if $A \subset B$,

$$
\begin{aligned}
\text{cap}(A) &= \lim_{m \to \infty} \sum_{x \in \partial C_m} P^x\{\tau_A < \xi_m\} \\
&\leq \lim_{m \to \infty} \sum_{x \in \partial C_m} P^x\{\tau_B < \xi_m\} \\
&= \text{cap}(B).
\end{aligned}
$$

For any finite A, B,

$$
\begin{aligned}
\text{cap}(A \cup B) &= \lim_{m \to \infty} \sum_{x \in \partial C_m} P^x\{\tau_{A \cup B} < \xi_m\} \\
&= \lim_{m \to \infty} \sum_{x \in \partial C_m} [P^x\{\tau_A < \xi_m\} + P^x\{\tau_B < \xi_m\}
\end{aligned}
$$

$$-P^x\{\tau_A < \xi_m, \tau_B < \xi_m\}]$$
$$\leq \quad \text{cap}(A) + \text{cap}(B) - \lim_{m\to\infty} \sum_{x\in\partial C_m} P^x\{\tau_{A\cap B} < \xi_m\}$$
$$= \quad \text{cap}(A) + \text{cap}(B) - \text{cap}(A\cap B). \quad \square$$

We now compute the capacity of the ball $\overline{C}_n$. If $A = \{0\}$, then

$$\text{cap}(A) = P\{\tau_0 = \infty\} = [G(0)]^{-1}.$$

Note that by Proposition 1.5.9,

$$P\{\tau_0 = \infty\} = P\{\tau_0 > \xi_m\}(1 + O(m^{2-d})).$$

Therefore by (2.1),

$$\sum_{x\in\partial C_m} P^x\{\tau_0 < \xi_m\} \quad = \quad P\{\tau_0 > \xi_m\}$$
$$= \quad [G(0)]^{-1}[1 - O(m^{2-d})].$$

If $\overline{C}_n \subset C_m$, then for $x \in \overline{C}_m \setminus \overline{C}_n$, again using Proposition 1.5.9,

$$P^x\{\tau_0 < \xi_m\} \quad = \quad P^x\{\xi_n < \xi_m\}P^x\{\tau_0 < \xi_m \mid \xi_n < \xi_m\}$$
$$= \quad P^x\{\xi_n < \xi_m\}[a_d G(0)^{-1}n^{2-d} + O(n^{1-d}) + O(m^{2-d})].$$

Therefore,

$$\sum_{x\in\partial C_m} P^x\{\xi_n < \xi_m\} =$$

$$\sum_{x\in\partial C_m} P^x\{\tau_0 < \xi_m\}[a_d G(0)^{-1}n^{2-d} + O(n^{1-d}) + O(m^{2-d})]^{-1}, \quad (2.15)$$

and if we let $m \to \infty$,

$$\text{cap}(\overline{C}_n) = a_d^{-1}n^{d-2} + O(n^{d-1}). \quad (2.16)$$

The capacity of a set is related to how likely a random walker that is close to A will hit A.

Proposition 2.2.2 *If $A \subset C_n$ and $x \in \partial C_{2n}$,*

$$P^x\{\tau_A < \infty\} \asymp n^{2-d}\text{cap}(A).$$

Proof. If $m > 3n, y \in \partial C_m$,

$$
\begin{aligned}
P^y\{\tau_A < \xi_m\} &= P^y\{\xi_{2n} < \xi_m\}P^y\{\tau_A < \xi_m \mid \xi_{2n} < \xi_m\} \\
&= P^y\{\xi_{2n} < \xi_m\}[P^y\{\tau_A < \infty \mid \xi_{2n} < \xi_m\} + O(m^{2-d})].
\end{aligned}
$$

By the Harnack principle (Theorem 1.7.6), if $x \in \partial C_{2n}$,

$$
P^x\{\tau_A < \infty\} \asymp P^y\{\tau_A < \infty \mid \xi_{2n} < \xi_m\}.
$$

Therefore,

$$
\begin{aligned}
\mathrm{cap}(A) &= \lim_{m\to\infty} \sum_{y\in\partial C_m} P^y\{\tau_A < \xi_m\} \\
&\asymp \lim_{m\to\infty} [\sum_{y\in\partial C_m} P^y\{\xi_{2n} < \xi_m\}]P^x\{\tau_A < \infty\} \\
&= \mathrm{cap}(\overline{C}_{2n})P^x\{\tau_A < \infty\}.
\end{aligned}
$$

The result then follows from (2.16). □

Proposition 2.2.3 *If* $A \subset C_{2n} \setminus C_n$, *then*

$$
P\{\tau_A < \infty\} \asymp n^{2-d}\mathrm{cap}(A).
$$

Proof. Let $A = A_+ \cup A_-$ where $A_+ = \{(z_1,\ldots,z_d) \in A : z_1 \geq 0\}, A_- = A \setminus A_+$. By the Harnack principle (Theorem 1.7.6), if $x \in \partial C_{4n}$,

$$
P\{\tau_{A_+} < \infty\} \asymp P^x\{\tau_{A_+} < \infty\},
$$

and hence by Proposition 2.2.2,

$$
P\{\tau_{A_+} < \infty\} \asymp n^{2-d}\mathrm{cap}(A_+).
$$

Similarly,

$$
P\{\tau_{A_-} < \infty\} \asymp n^{2-d}\mathrm{cap}(A_-).
$$

Then, using Proposition 2.2.1,

$$
\begin{aligned}
P\{\tau_A < \infty\} &\leq P\{\tau_{A_+} < \infty\} + P\{\tau_{A_-} < \infty\} \\
&\asymp n^{2-d}(\mathrm{cap}(A_+) + \mathrm{cap}(A_-)) \\
&\asymp n^{2-d}\mathrm{cap}(A). \\
P\{\tau_A < \infty\} &\geq \sup\{P\{\tau_{A_+} < \infty\}, P\{\tau_{A_-} < \infty\}\} \\
&\asymp n^{2-d}\sup\{\mathrm{cap}(A_+), \mathrm{cap}(A_-)\} \\
&\asymp n^{2-d}\mathrm{cap}(A). □
\end{aligned}
$$

Now suppose $A \subset Z^d$ is infinite. We call A a *recurrent* set if

$$P\{S_j \in A \text{ infinitely often}\} = 1,$$

and a *transient* set if

$$P\{S_j \in A \text{ infinitely often}\} = 0.$$

Proposition 2.2.4 *Every set $A \subset Z^d$ is either recurrent or transient.*

Proof. Let $f(x) = P^x\{S_j \in A \text{ i. o.}\}$. Then f is a bounded harmonic function, and hence (see Exercise 1.4.10), f is constant, say $f \equiv k$. Let V be the event $\{S_j \in A \text{ i. o.}\}$. Since $f \equiv k$, $P^0(V \mid \mathcal{F}_n) = k$ for each n, i.e. V is independent of $\mathcal{F}_n$. Therefore V is a tail event, and by the Kolmogorov 0-1 Law, $P^0(V) = 0$ or 1. $\square$

Theorem 2.2.5 (Wiener's Test) *Suppose $A \subset Z^d$ $(d \geq 3)$ and let*

$$A_n = \{z \in A : 2^n \leq |z| < 2^{n+1}\}.$$

Then A is a recurrent set if and only if

$$\sum_{n=0}^{\infty} \frac{\text{cap}(A_n)}{2^{n(d-2)}} = \infty.$$

Proof. Let I_n be the indicator function of the event $V_n = \{\tau_{A_n} < \infty\}$. Then since each A_n is finite,

$$P\{S_j \in A \text{ i. o.}\} = P\{I_n = 1 \text{ i. o.}\}.$$

By Proposition 2.2.3,

$$P(V_n) \asymp \text{cap}(A_n) 2^{(2-d)n}.$$

Therefore,

$$\sum_{n=0}^{\infty} P(V_n) = \infty \Longleftrightarrow \sum_{n=0}^{\infty} \frac{\text{cap}(A_n)}{2^{n(d-2)}} = \infty.$$

Suppose $\sum P(V_n) < \infty$. Then by the Borel-Cantelli Lemma, $P\{I_n = 1 \text{ i. o.}\} = 0$ and hence A is transient. Suppose then that $\sum P(V_n) = \infty$. Then either $\sum_{n=0}^{\infty} P(V_{2n}) = \infty$ or $\sum_{n=0}^{\infty} P(V_{2n+1}) = \infty$. Assume the former (a similar argument works in the latter case). Let $m \geq n+2, \tau_n = \tau_{A_n}$, and consider

$$V_n \cap V_m = \{\tau_n < \tau_m < \infty\} \cup \{\tau_m < \tau_n < \infty\}.$$

By the Harnack principle and the strong Markov property,

$$P\{\tau_n < \tau_m < \infty\} \;\leq\; P\{\tau_n < \infty; S_j \in A_m \text{ for some } j > \tau_n\}$$
$$\asymp\; P(V_n)P(V_m).$$

Similarly, using Propositions 2.2.2 and 2.2.3,

$$P\{\tau_m < \tau_n < \infty\} \;\leq\; P\{\tau_m < \infty; S_j \in A_n \text{ for some } j > \tau_m\}$$
$$\leq\; cP(V_n)P(V_m).$$

Therefore, for some $c < \infty$, if $0 \leq n \leq m - 2$,

$$P(V_n \cap V_m) \leq c\, P(V_n)P(V_m). \tag{2.17}$$

Let $J_n = \sum_{j=0}^{n} I_{2j}$ and for $\epsilon > 0$, let F_ϵ be the indicator function of $\{J_n \geq \epsilon E(J_n)\}$. Then,

$$[E(F_\epsilon J_n)]^2 \leq E(F_\epsilon^2)E(J_n^2).$$

But by (2.17),

$$E(J_n^2) \leq c\,[E(J_n)]^2,$$

and

$$E(F_\epsilon J_n) = E(J_n) - E(J_n\, I\{J_n < \epsilon E(J_n)\}) \geq (1 - \epsilon)E(J_n).$$

Therefore,

$$P\{J_n \geq \epsilon E(J_n)\} = E(F_\epsilon) = E(F_\epsilon^2) \geq \frac{1}{c}(1 - \epsilon)^2.$$

If we let $\epsilon = \frac{1}{2}$ and let $n \to \infty$, then $E(J_n) \to \infty$ and hence

$$P\{J_\infty = \infty\} \geq \frac{1}{2c} > 0,$$

which implies by Proposition 2.2.4 that A is recurrent.   □

Let V_A be the number of visits to A, i.e.,

$$V_A = \sum_{j=0}^{\infty} \sum_{x \in A} I\{S_j = x\}.$$

Then it is easy to see that

$$E(V_A) = \sum_{x \in A} \sum_{j=0}^{\infty} P\{S_j = x\} = \sum_{x \in A} G(x).$$

If $E(V_A) < \infty$, then $V_A < \infty$ almost surely and A is transient. The converse, however, is not true. For example, suppose $\alpha \in (\frac{d-2}{d}, 1)$, A_n is a "ball" of radius $2^{\alpha n}$ contained in $\{z : 2^n \leq |z| < 2^{n+1}\}$, and $A = \cup_{n=1}^{\infty} A_n$. Then by (2.16), $\text{cap}(A_n) \asymp 2^{\alpha(d-2)n}$ and hence by Wiener's Test (Theorem 2.2.5), A is transient. However, by Theorem 1.5.4,

$$E(V_A) = \sum_{n=1}^{\infty} \sum_{y \in A_n} G(y)$$

$$\geq c \sum_{n=1}^{\infty} 2^{\alpha n d} 2^{(2-d)n} = \infty.$$

Exercise 2.2.6 *Let $A_k \subset Z^d$ be the set*

$$A_k = \{(z_1, \ldots, z_d) \in Z^d : z_1 = z_2 = \cdots = z_k = 0\}.$$

Show that A_k is a recurrent set if and only if $d - k \leq 2$.

If $f : \overline{A} \to R$, we define the "(outward) normal derivative" $\nabla_N f(x)$ for $x \in A$ by

$$\nabla_N f(x) = \frac{1}{2d} \sum_{|e|=1, x+e \in \partial A} (f(x+e) - f(x)). \tag{2.18}$$

Let $\overline{\tau} = \inf\{j \geq 0 : S_j \in A\}$ and let

$$g(x) = g_A(x) = P^x\{\overline{\tau} = \infty\}.$$

Then g is harmonic for $x \in A^c$ and $g \equiv 0$ on A. If A is finite, then $\lim_{|x| \to \infty} g(x) = 1$.

Exercise 2.2.7 *If A is a finite set, then*

$$H_A(x) = \frac{\nabla_N g_A(x)}{\sum_{y \in A} \nabla_N g_A(y)}.$$

2.3 Capacity, Two Dimensions

In this section we give the two dimensional analogue of the capacity. Let $A \subset Z^2$ be a finite set, $z \in A$, and as in the previous section $\xi_n = \tau_{\partial C_n}$. Let $a(x)$ be the potential kernel as defined in Section 1.6.

Lemma 2.3.1 *For every $z \in A$, $x \notin A$,*

$$\lim_{m \to \infty} (\frac{2}{\pi} \ln m) P^x \{\xi_m < \tau_A\} = a(x - z) - \sum_{y \in A} H_A(x, y) a(y - z).$$

Proof. Assume $A \subset C_n$, $m > |x|, n$. Let $\eta = \xi_m \wedge \tau_A$. If $S_0 = x$, then $M_j = a(S_{j \wedge \eta} - z)$ is a bounded martingale, and by the optional sampling theorem,

$$
\begin{aligned}
a(x - z) &= E^x(M_\eta) \\
&= \sum_{y \in A} P^x\{S_\eta = y\}a(y - z) \\
&\quad + P^x\{\xi_m < \tau_A\}E^x(a(S_\eta - z) \mid \xi_m < \tau_A).
\end{aligned}
$$

Since $|z| < n$, Theorem 1.6.2 gives

$$
E^x(a(S_\eta - z) \mid \xi_m < \tau_A) = \frac{2}{\pi}\ln m + k + no(m^{-1}).
$$

If we take the limit as $m \to \infty$, we get the lemma. $\square$

One consequence of Lemma 2.3.1 is that the function on A^c,

$$
g_A(x) = g_{z,A}(x) = a(x - z) - \sum_{y \in A} H_A(x, y)a(y - z)
$$

is independent of $z \in A$. For ease we will assume $0 \in A$, and let

$$
\tilde{H}_A(x, y) = \begin{cases} H_A(x, y), & x \in A^c, \\ \delta(x - y), & x \in A, \end{cases}
$$

$$
g_A(x) = a(x) - \sum_{y \in A} \tilde{H}_A(x, y)a(y).
$$

Proposition 2.3.2 *Suppose $0 \in A \subset C_n$. Then if $g(x) = g_A(x)$,*
(a) $g(y) = 0$, $y \in A$;
(b) $\Delta g(x) = 0$, $x \in A^c$;
(c) As $|x| \to \infty$,

$$
g(x) = \frac{2}{\pi}\ln|x| + k - \sum_{y \in A} H_A(y)a(y) + c_n O\left(\frac{\ln|x|}{|x|}\right);
$$

(d) If $y \in A$,

$$
H_A(y) = \frac{\nabla_N g(y)}{\sum_{z \in A}\nabla_N g(z)}.
$$

Proof. (a) is immediate and (b) follows from the fact that $a(x)$ and $\tilde{H}_A(x, y)$ are harmonic for $x \in A^c$. To prove (c), note that Theorem 1.6.2 gives

$$
a(x) = \frac{2}{\pi}\ln|x| + k + o(|x|^{-1}),
$$

while Theorem 2.1.3 gives

$$H_A(x, y) = H_A(y)[1 + O(\frac{n}{|x|} \ln \frac{|x|}{n})].$$

Part (d) follows from Theorem 2.1.3 and Lemma 2.3.1. $\square$

Proposition 2.3.3 *Suppose $A \subset C_n$ and $h : Z^2 \to R$ is a function satisfying*

(a) $h(x) = 0$, $x \in A$;

(b) $\Delta h(x) = 0$, $x \in A^c$;

(c) $\limsup_{|x| \to \infty} \dfrac{|h(x)|}{\ln |x|} < \infty.$

Then $h(x) = Cg_A(x)$ for some $C \in R$.

Proof. Assume for ease that $0 \in A$. As in Lemma 2.3.1, let $\eta = \xi_m \wedge \tau_A$, $M_j = h(S_{j \wedge \eta})$. Then by optional sampling, for $x \in C_m \setminus A$,

$$h(x) = E^x(M_\eta) = \sum_{y \in \partial C_m} P^x\{S_\eta = y\}h(y).$$

By an argument identical to the proof of Lemma 2.1.2, if $y \in \partial C_m$,

$$P^x\{S_\eta = y\} = P^x\{\xi_m < \tau_A\}H_{\partial C_m}(0, y)[1 + O(\frac{|x|}{m} \ln \frac{m}{|x|})].$$

Therefore, since $h(y) = O(\ln |y|)$,

$$h(x) = P^x\{\xi_m < \tau_A\}[\sum_{y \in A} H_{\partial C_m}(0, y)h(y)] + O(\frac{|x|}{m} \ln^2 \frac{m}{|x|}),$$

and therefore by Lemma 2.3.1, if we let $m \to \infty$,

$$h(x) = [a(x) - \sum_{y \in A} H_A(x, y)a(y)]C,$$

where

$$C = \lim_{m \to \infty} (\frac{2}{\pi} \ln m)^{-1} \sum_{y \in \partial C_m} P^x\{S_\eta = y\}H_{\partial C_m}(0, y)h(y). \quad \square$$

If $0 \in A$, we define the *capacity* or *Robin's constant* of the set A to the the number

$$\text{cap}(A) = \sum_{y \in A} H_A(y)a(y) - k,$$

so that

$$g_A(x) = \frac{2}{\pi} \ln |x| - \text{cap}(A) + c_n O(\frac{\ln |x|}{|x|}).$$

The capacity is translation invariant, and we define $\text{cap}(A)$ for sets not containing 0 by first translating to a set containing 0. As one can guess from the choice of terminology, the capacity of a two dimensional set has many of the properties that the capacity in the last section has.

Proposition 2.3.4 *If A and B are finite subsets of Z^2,*
  (a) If $A \subset B$, then $\text{cap}(A) \leq \text{cap}(B)$;
  (b) $\text{cap}(A) + \text{cap}(B) \geq \text{cap}(A \cup B) + \text{cap}(A \cap B)$.

  Proof. Assume for ease that $0 \in A$. Let $x \notin A \cup B$. By Lemma 2.3.1,

$$\lim_{m \to \infty} \frac{2}{\pi} (\ln m) P^x \{\xi_m < \tau_A\} = a(x) + k - \text{cap}(A).$$

If $A \subset B$,

$$P^x \{\xi_m < \tau_A\} > P^x \{\xi_m > \tau_B\},$$

and hence $\text{cap}(A) \leq \text{cap}(B)$. In general,

$$\begin{aligned}
P^x \{\xi_m > \tau_{A \cup B}\} &= P^x \{\xi_m > \tau_A\} + P^x \{\xi_m > \tau_B\} \\
&\quad - P^x \{\xi_m > \tau_A, \xi_m > \tau_B\} \\
&\leq P^x \{\xi_m > \tau_A\} + P^x \{\xi_m > \tau_B\} - P^x \{\xi_m > \tau_{A \cap B}\},
\end{aligned}$$

which gives (b). □
  The capacity of a singleton set $\{x\}$ is $-k$. Suppose $A = C_n$. Then by Exercise 1.6.8, if $x \notin C_n, m > |x|$,

$$P^x \{\tau_A > \xi_m\} = \frac{\ln |x| - \ln n + O(n^{-1})}{\ln m - \ln n}.$$

Therefore, by Lemma 2.3.1, if $x \notin C_n$,

$$g_A(x) = \frac{2}{\pi} \{\ln |x| - \ln n\} + O(n^{-1}).$$

By Proposition 2.3.2(c),

$$g_A(x) = \frac{2}{\pi} \ln |x| - \text{cap}(C_n) + c_n O(\frac{\ln |x|}{|x|}).$$

Since this holds for all $x \notin C_n$, we can let $|x| \to \infty$ and get

$$\text{cap}(C_n) = \frac{2}{\pi} \ln n + O(n^{-1}). \tag{2.19}$$

We call a set $A \subset Z^d$ *connected* if every two points in A can be joined by a random walk path in A, i.e., if for every $x, y \in A$, $\exists x_0, \ldots, x_m \in A$ with $x = x_0, y = x_m, |x_j - x_{j-1}| = 1$ for $1 \leq j \leq m$. We define the *radius* of A, rad(A), to be the smallest integer n such that $A \subset C_n$.

Lemma 2.3.5 *If A is a connected subset of Z^2 containing 0 of radius n, then*

$$\frac{2}{\pi} \ln n - O(1) \leq \text{cap}(A) \leq \frac{2}{\pi} \ln n + O(n^{-1}).$$

Proof. The right hand inequality follows from (2.19) and Proposition 2.3.4(a). To prove the other inequality, find a subset B of A such that for each $j = 1, 2, \ldots, n$, there exists exactly one point $x \in B$ with $j - 1 \leq |x| < j$. Since A is connected with radius n, one can always find such a subset (although the subset might not be connected). Again by Proposition 2.3.4(a), it suffices to prove the left hand inequality for B. Let $m > 2n$. By Proposition 1.6.7, as $|m| \to \infty$, if $x, y \in C_n$,

$$G_{C_m}(x, y) = \frac{2}{\pi} \{\ln m - \ln |x - y| + O(1)\}. \tag{2.20}$$

(Here we use the inequality

$$G_{C_{m-2n}}(0, y - x) \leq G_{C_m}(x, y) \leq G_{C_{m+2n}}(0, y - x).)$$

Let V_m be the number of visits to B before leaving C_m, i.e.,

$$V_m = \sum_{j=0}^{\xi_m} I\{S_j \in B\}.$$

By (2.20), for each $x \in \overline{C}_n$,

$$E^x(V_m) = \sum_{y \in B} G_{C_m}(x, y) \geq \frac{2n}{\pi} \{\ln m - \ln n + O(1)\}. \tag{2.21}$$

Moreover, if $x \in \overline{C}_n$, there exist at most $2j$ points in B at a distance less than or equal to $j - 1$ from x. Therefore,

$$E^x(V_m) = \sum_{y \in B} G_{C_m}(x, y) \leq \frac{2}{\pi} \{n \ln m - 2 \sum_{j=2}^{n/2} \ln j + nO(1)\}$$

$$= \frac{2n}{\pi} \{\ln m - \ln n + O(1)\}. \tag{2.22}$$

If $x \in \partial C_n$,

$$E^x(V_m) = P^x\{\tau_B < \xi_m\}E^x(V_m \mid \tau_B < \xi_m).$$

Therefore by (2.21) and (2.22), if $x \in \partial C_n$,

$$P^x\{\tau_B > \xi_m\} = O((\ln \frac{m}{n})^{-1}). \tag{2.23}$$

Hence if $z \in C_m \setminus \overline{C}_n$, by Exercise 1.6.8,

$$
\begin{aligned}
P^z\{\tau_B < \xi_m\} &= P^z\{\xi_n < \xi_m\}[1 - O((\ln \frac{m}{n})^{-1})] \\
&= \frac{\ln m - \ln |z| + O(n^{-1})}{\ln m - \ln n} - O((\ln \frac{m}{n})^{-1}),
\end{aligned}
$$

and

$$P^z\{\tau_B > \xi_m\} = \frac{\ln |z| - \ln n + O(n^{-1})}{\ln m - \ln n} + O((\ln \frac{m}{n})^{-1}).$$

Therefore by Lemma 2.3.1,

$$g_B(z) = \frac{2}{\pi}\{\ln |z| - \ln n\} + O(1).$$

Again we use Proposition 2.3.2(c) to give

$$g_B(z) = \frac{2}{\pi}\ln |z| - \mathrm{cap}(B) + c_n O(\frac{\ln |z|}{|z|}),$$

and letting $|z| \to \infty$,

$$\mathrm{cap}(B) = \frac{2}{\pi}\ln n + O(1). \quad \square$$

2.4   Example: Line Segment

In this section we will study in detail the examples of a line and a line segment in Z^d. In the process we will illustrate techniques which are used to relate various escape probabilities and harmonic measure. Let $\langle n \rangle = (n, 0, \ldots, 0)$ and

$$
\begin{aligned}
U &= \{\langle n \rangle : n \in Z\}, \\
U^+ &= \{\langle n \rangle : n \geq 0\}, \\
U^- &= \{\langle n \rangle : n < 0\}, \\
U_n &= \{\langle j \rangle : -n < j < n\}, \\
U_n^+ &= U^+ \cap U_n.
\end{aligned}
$$

We have already seen (see exercise 2.2.6) that U is a transient set if and only if $d \geq 4$. We will be most interested here in $d = 2, 3$. We start with a very useful proposition which relates escape probabilities and the Green's function.

Proposition 2.4.1 *Suppose $A \subset Z^d$, and*

$$\begin{aligned} \tau &= \inf\{j \geq 1 : S_j \in A\}, \\ \bar{\tau} &= \inf\{j \geq 0 : S_j \in A\}. \end{aligned}$$

(a) If $n < \infty$, then

$$P^x\{\bar{\tau} \leq n\} \geq \sum_{y \in A} G_n(x, y)P^y\{\tau > n\}.$$

(b) If $\lambda < 1$, and T is a killing time with rate $1 - \lambda$, then

$$P^x\{\bar{\tau} \leq T\} = \sum_{y \in A} G_\lambda(x, y)P^y\{\tau > T\}.$$

(c) If $A \subset B$ and $\eta = \inf\{j \geq 0 : S_j \notin B\}$, then

$$P^x\{\bar{\tau} \leq \eta\} = \sum_{y \in A} G_B(x, y)P^y\{\tau > \eta\}.$$

(d) If $d \geq 3$ and A is a transient set, then

$$P^x\{\bar{\tau} < \infty\} = \sum_{y \in A} G(x, y)\mathrm{Es}_A(y).$$

Proof. We will prove (c); the other proofs are similar. The proof uses a technique sometimes called a "last-exit decomposition." Let

$$\sigma = \sup\{j : S_j \in A, j \leq \eta\}.$$

Then,

$$\begin{aligned} P^x\{\bar{\tau} \leq \eta\} &= \sum_{j=0}^{\infty} P^x\{\sigma = j\} \\ &= \sum_{y \in A} \sum_{j=0}^{\infty} P^x\{\sigma = j, S_j = y\}. \end{aligned}$$

By the Markov property,

$$\begin{aligned} P^x\{\sigma = j, S_j = y\} &= P^x\{S_j = y; j \leq \eta; S_k \notin A, j < k \leq \eta\} \\ &= P^x\{S_j = y; j \leq \eta\}P^y\{\tau > \eta\}. \end{aligned}$$

Therefore,

$$
\begin{aligned}
P^x\{\bar{\tau} \leq \eta\} &= \sum_{y \in A} P^y\{\tau > \eta\} \sum_{j=0}^{\infty} P^x\{S_j = y, j \leq \eta\} \\
&= \sum_{y \in A} G_B(x, y) P^y\{\tau > \eta\}. \quad \square
\end{aligned}
$$

If one wants to analyze walks which take only a finite number of steps, it can be easier to first consider walks with a killing rate $1 - \lambda$ rather than with with a fixed number of steps, since there is equality rather than inequality in the above proposition. If $\lambda_n = 1 - \frac{1}{n}$, then the random walk on the average takes n steps, so one would hope that results about walks with rate $1 - \lambda_n$ could be used to get results about walks of length n, and vice versa. This is in fact true, and the mathematical tool used is Tauberian theory. If p_n is any sequence of nonnegative numbers, then the *generating function* of p_n, $R(\lambda)$, is defined by

$$
R(\lambda) = \sum_{n=0}^{\infty} \lambda^n p_n.
$$

As an example, let $p_n = P^x\{\tau > n\}$. Then

$$
\begin{aligned}
P^x\{\tau > T\} &= \sum_{n=0}^{\infty} P\{T = n\} P^x\{\tau > n\} \\
&= (1 - \lambda) R(\lambda). \tag{2.24}
\end{aligned}
$$

We state a Tauberian theorem which relates p_n to its generating function. The proof can be found in [25, Chapter XIII]. We say a function $L : [0, \infty) \rightarrow [0, \infty)$ is *slowly varying (at infinity)* if for every $t \in (0, \infty)$, $L(tx) \sim L(x)$ as $|x| \rightarrow \infty$.

Theorem 2.4.2 *Suppose p_n is a sequence of nonnegative real numbers; L a slowly varying function; and*

$$
R(\lambda) = \sum_{n=0}^{\infty} \lambda^n p_n.
$$

Then if $\alpha \in [0, \infty)$, the following are equivalent:
  (a) as $\lambda \rightarrow 1-$,

$$
R(\lambda) \sim (1 - \lambda)^{-\alpha} L(\frac{1}{1 - \lambda});
$$

(b) as $n \to \infty$,

$$\sum_{j=0}^{n-1} p_j \sim [\Gamma(\alpha+1)]^{-1} n^\alpha L(n).$$

Moreover, if the p_n are monotone and $\alpha > 0$, the following is equivalent to (a) and (b):
(c) as $n \to \infty$,

$$p_n \sim [\Gamma(\alpha)]^{-1} n^{\alpha-1} L(n).$$

In some of the examples we will consider, we will know the behavior of p_n or $R(\lambda)$ only up to a multiplicative constant. It will be useful to have a form of the above theorem which handles this case.

Theorem 2.4.3 *Let p_n, L, R, α be as in Theorem 2.4.2. Then the following are equivalent:*
(a) there exists $0 < b_1 < b_2 < \infty$ with

$$b_1(1-\lambda)^{-\alpha} L((1-\lambda)^{-1}) \le R(\lambda) \le b_2(1-\lambda)^{-\alpha} L((1-\lambda)^{-1});$$

(b) there exist $0 < \beta_1 < \beta_2 < \infty$ with

$$\beta_1 n^\alpha L(n) \le \sum_{j=0}^{n-1} p_j \le \beta_2 n^\alpha L(n).$$

Moreover, if the p_n are monotone and $\alpha > 0$, the following is equivalent to (a) and (b):
(c) there exist $0 < \delta_1 < \delta_2 < \infty$ with

$$\delta_1 n^{\alpha-1} L(n) \le p_n \le \delta_2 n^{\alpha-1} L(n).$$

Proof. The fact that (b) implies (a) follows from Theorem 2.4.2 by comparing $R(\lambda)$ to the generating function for $\beta_i n^\alpha L(n)$. Assume (a) holds. It suffices to prove (b) for n sufficiently large. If $\lambda = 1 - \frac{1}{n}$,

$$
\begin{aligned}
\sum_{j=0}^{n-1} p_j &\le \lambda^{-n} \sum_{j=0}^{n-1} \lambda^j p_j \\
&\le cR(\lambda) \\
&\le cb_2 n^\alpha L(n). \quad (2.25)
\end{aligned}
$$

To prove the other inequality let $a > 0$ and $\lambda = 1 - \frac{a}{n}$. Then, using (2.25),

$$
\begin{aligned}
b_1 a^{-\alpha} n^\alpha L(n/a) \;\leq\; & R(\lambda) \\
=\; & \sum_{j=0}^{n-1} \lambda^j p_j + \sum_{j=n}^{\infty} \lambda^j p_j \\
\leq\; & \sum_{j=0}^{n-1} \lambda^j p_j + \sum_{k=1}^{\infty} \lambda^{kn} \sum_{j=nk}^{n(k+1)} p_j \\
\leq\; & \sum_{j=0}^{n-1} \lambda^j p_j + \sum_{k=1}^{\infty} e^{-ak}[cb_2(k+1)^\alpha n^\alpha L((k+1)n)].
\end{aligned}
$$

If L is slowly varying, then $L(kn) \leq c_L k L(n)$ [25, (9.9)]. Also for n large, $L(n/a) \geq \frac{1}{2} L(n)$, hence

$$
\sum_{j=0}^{n-1} \lambda^j p_j \geq n^\alpha L(n)\{\tfrac{1}{2}b_1 a^{-\alpha} - c_L b_2 \sum_{k=1}^{\infty} e^{-ak}(k+1)^{\alpha+1}\}.
$$

If we choose a sufficiently large (depending on b_1, b_2, α, L) we can make the coefficient on the right positive and obtain (b).

By summing, it is easy to see that if $\alpha > 0$, (c) implies (b). Suppose (b) holds, $\alpha > 0$, and p_n is monotone. Assume p_n is a decreasing sequence (a similar argument holds if p_n is increasing). Then

$$
p_n \leq n^{-1} \sum_{j=0}^{n-1} p_j \leq \beta_2 n^{\alpha-1} L(n).
$$

To prove the other direction, let $a > 0$ and note that

$$
\sum_{j=0}^{an-1} p_j \;\geq\; \beta_1(an)^\alpha L(an),
$$

$$
\sum_{j=0}^{n-1} p_j \;\leq\; \beta_2 n^\alpha L(n),
$$

and hence,

$$
(a-1)n p_n \geq \sum_{j=n}^{an-1} p_j \geq \beta_1(an)^\alpha L(an) - \beta_2 n^\alpha L(n).
$$

Choose a sufficiently large so that $\beta_1 a^\alpha \geq 4\beta_2$ and then N sufficiently large so that $L(an) \geq \frac{1}{2}L(n)$ for $n \geq N$. Then for $n \geq N$,

$$p_n \geq (a-1)^{-1}\beta_2 L(n)n^{\alpha-1}. \quad \square$$

To illustrate the usefulness of the above theorems we will return to studying the straight line.

Lemma 2.4.4

$$P\{S_j \in U\} \sim (\frac{d}{2\pi j})^{(d-1)/2}.$$

Proof.

$$
\begin{aligned}
P\{S_j \in U\} &= \sum_{n=-\infty}^{\infty} p(j, \langle n \rangle) \\
&= \sum_{n \leftrightarrow j} [\overline{p}(j, \langle n \rangle) + E(j, \langle n \rangle)].
\end{aligned}
$$

By (1.10) and (1.11),

$$
\begin{aligned}
|\sum_{n=-\infty}^{\infty} E(j, \langle n \rangle)| &\leq \sum_{|n| \leq \sqrt{j}} |E(j, \langle n \rangle)| + \sum_{n > \sqrt{j}} |E(j, \langle n \rangle)| \\
&= \sum_{|n| \leq \sqrt{j}} O(j^{-(d+2)/2}) + \sum_{n > \sqrt{j}} n^{-2}O(j^{-d/2}) \\
&= O(j^{-(d+1)/2}),
\end{aligned}
$$

so we only need to estimate the dominant term. But,

$$
\begin{aligned}
\sum_{n \leftrightarrow j} \overline{p}(j, \langle n \rangle) &= \sum_{n \leftrightarrow j} 2(\frac{d}{2\pi j})^{d/2} \exp\{-\frac{dn^2}{2j}\} \\
&\sim \int_{-\infty}^{\infty} (\frac{d}{2\pi j})^{d/2} \exp\{-\frac{dx^2}{2j}\}dx = (\frac{d}{2\pi j})^{(d-1)/2}. \quad \square
\end{aligned}
$$

Since

$$\sum_{x \in U} G_\lambda(x) = \sum_{j=0}^{\infty} \lambda^j P\{S_j \in U\},$$

it follows immediately from Theorem 2.4.2 and Lemma 2.4.4 that as $\lambda \to 1-$,

$$\sum_{x \in U} G_\lambda(x) \sim \begin{cases} (1-\lambda)^{-1/2}, & d = 2, \\ \frac{3}{2\pi} \ln \frac{1}{1-\lambda}, & d = 3. \end{cases} \quad (2.26)$$

Since $P^x\{\tau > T\} = P^0\{\tau > T\}$ for each $x \in U$, it follows from Proposition 2.4.1(b) that

$$P\{\tau > T\} \sim \begin{cases} (1-\lambda)^{1/2}, & d = 2, \\ \frac{2\pi}{3}(\ln \frac{1}{1-\lambda})^{-1}, & d = 3. \end{cases} \tag{2.27}$$

Another application of Theorem 2.4.2, using (2.24), gives

$$P\{\tau_U > n\} \sim \begin{cases} (\pi n)^{-1/2}, & d = 2, \\ \frac{2\pi}{3}(\ln n)^{-1}, & d = 3. \end{cases}$$

We now consider the line segment U_n. If $d \geq 4$ and $x \in U_n$, then $\mathrm{Es}_{U_n}(x) \geq \mathrm{Es}_U(x) > 0$, and hence $\mathrm{cap}(U_n) \asymp n$. If $d = 2$, it follows from Theorem 2.3.5 that

$$\mathrm{cap}(U_n) = \frac{2}{\pi}\ln n + O(1).$$

Proposition 2.4.5 *If* $|m| < n$, $\eta = \tau_{U_n}$, *and* $\xi = \xi_{2n}$,

$$c_1 n^{-1} \leq P^{\langle m \rangle}\{\xi < \eta\} \leq c_2(n - |m|)^{-1}, \quad d = 2;$$

$$c_1(\ln n)^{-1} \leq P^{\langle m \rangle}\{\xi < \eta\} \leq c_2(\ln(n - |m|))^{-1}, \quad d = 3.$$

Proof. Let $g(n) = P\{\xi < \eta\}$. Then it is easy to see that for $|m| < n$,

$$g(n + |m|) \leq P^{\langle m \rangle}\{\xi < \eta\} \leq g(n - |m|),$$

and hence it suffices to prove the proposition for $m = 0$. By Proposition 2.4.1(c),

$$\begin{aligned} 1 &= \sum_{|m|<n} G_{C_{2n}}(0, \langle m \rangle)P^{\langle m \rangle}\{\xi < \eta\} \\ &\geq g(2n) \sum_{|m|<n} G_{C_{2n}}(0, \langle m \rangle). \end{aligned}$$

Therefore by Proposition 1.6.7 and Proposition 1.5.9,

$$g(n) \leq [\sum_{|m|<n/2} G_{C_n}(0, \langle m \rangle)]^{-1} \leq \begin{cases} cn^{-1}, & d = 2, \\ c(\ln n)^{-1}, & d = 3. \end{cases}$$

To get the lower bound, we first note that if $W = U_n \setminus U_{n/2}$, then an application of the Harnack principle gives

$$P\{\xi < \tau_W\} \geq c.$$

But, by examining the proof of Proposition 2.4.1(c), we see

$$\sum_{|m|<n/2} G_{C_{2n}}(0, \langle m \rangle) P^{\langle m \rangle}\{\xi < \eta\} \;=\; P\{|\sigma| < n/2\}$$

$$\geq \;\; P\{\xi < \tau_W\} \;\geq\; c.$$

Therefore

$$c \leq g\left(\frac{n}{2}\right) \sum_{|m|<n/2} G_{C_{2n}}(0, \langle m \rangle),$$

and the result follows again from Propositions 1.6.7 and 1.5.9. □

It follows from the proposition and Exercise 2.1.4 that

$$\frac{c_1}{n} \leq H_{U_n}(\langle m \rangle) \leq \frac{c_2 \ln n}{n \ln(n - |m|)}, \quad d = 3. \tag{2.28}$$

For $d = 2$, note by (2.1) and (2.23),

$$\sum_{x \in U_n} P^x\{\xi_{2n} < \tau_{U_n}\} \;=\; \sum_{y \in \partial C_{2n}} P^y\{\xi_{2n} > \tau_{U_n}\}$$

$$\asymp \;\; \sum_{y \in \partial C_{2n}} P^y\{\xi_{2n} > \xi_n\}.$$

But by Proposition 1.6.7,

$$P^y\{\xi_{2n} > \xi_n\} \asymp P^y\{\xi_{2n} > \tau_0\}(\ln n),$$

and hence

$$\sum_{y \in \partial C_{2n}} P^y\{\xi_{2n} > \xi_n\} \;\asymp\; (\ln n) \sum_{y \in \partial C_{2n}} P^y\{\xi_{2n} > \tau_0\}$$

$$= \;\; (\ln n) P\{\xi_{2n} < \tau_0\} \asymp 1. \tag{2.29}$$

Therefore,

$$\frac{c_1}{n} \leq H_{U_n}(\langle m \rangle) \leq \frac{c_2}{n - |m|}, \quad d = 2. \tag{2.30}$$

These estimates on the harmonic measure are good except for $|m|$ near n.

We now consider the half-line U^+. Assume we have a random walk with killing rate $1 - \lambda$ and let V^+, V^- be the events $\{S_j \notin U^+, 0 < j \leq T\}, \{S_j \notin U^-, 0 < j \leq T\}$, respectively. Then by (2.27),

$$P(V^+ \cap V^-) \sim \begin{cases} (1 - \lambda)^{1/2}, & d = 2, \\ \frac{2\pi}{3}[\ln(\frac{1}{1-\lambda})]^{-1}, & d = 3. \end{cases}$$

This fact alone does not allow us to estimate $P(V^+)$ and $P(V^-)$. However, we have the following (very nonintuitive) fact.

Proposition 2.4.6 *If $S_0 = 0$, V^+ and V^- are independent events.*

Proof. Let $W = (V^-)^c$. It suffices to prove

$$P(W \cap V^+) = P(W)P(V^+).$$

Let

$$\sigma = \sup\{n : \exists j, 1 \leq j \leq T, \text{ with } S_j = \langle n \rangle\},$$

$$\eta = \sup\{j \leq T : S_j = \langle \sigma \rangle\}.$$

Then,

$$
\begin{aligned}
P(W \cap V^+) &= \sum_{n=1}^{\infty} P\{\sigma = -n\} \\
&= \sum_{n=1}^{\infty} \sum_{j=1}^{\infty} P\{\sigma = -n, \eta = j\}.
\end{aligned}
\tag{2.31}
$$

For any n, j,

$$
\begin{aligned}
P\{\sigma = -n, \eta = j\} &= P\{S_j = \langle -n \rangle; S_i \neq \langle m \rangle, 1 \leq i < j, -n < m < \infty; \\
&\qquad T \geq j; S_i \neq \langle m \rangle, j < i \leq T, -n \leq m < \infty\} \\
&= P(V^+)P\{S_j = \langle -n \rangle; S_i \neq \langle m \rangle, 1 \leq i < j, \\
&\qquad -n < m < \infty; T \geq j\}.
\end{aligned}
\tag{2.32}
$$

If we reverse paths and translate we see that

$$
\begin{aligned}
&P\{S_j = \langle -n \rangle; S_i \neq \langle m \rangle, 1 \leq i < j, -n < m < \infty; T \geq j\} \\
&= P^{\langle -n \rangle}\{S_j = 0; S_i \neq \langle m \rangle, 1 \leq i < j, -n < m < \infty; T \geq j\} \\
&= P\{S_j = \langle n \rangle; S_i \neq \langle m \rangle, 1 \leq i < j, 0 < m < \infty; T \geq j\}.
\end{aligned}
\tag{2.33}
$$

Now let

$$\rho = \inf\{1 \leq j \leq T : S_j \in U^+ \setminus \{0\}\}.$$

Then by symmetry, $P\{\rho < \infty\} = P(W)$. Therefore,

$$
\begin{aligned}
P(W) &= \sum_{j=1}^{\infty} P\{\rho = j\} \\
&= \sum_{j=1}^{\infty} \sum_{n=1}^{\infty} P\{\rho = j, S_j = \langle n \rangle\}.
\end{aligned}
\tag{2.34}
$$

But,

$$P\{\rho = j, S_j = \langle n \rangle\} =$$
$$P\{S_j = \langle n \rangle; S_i \neq \langle m \rangle, 1 \leq i < j, 0 < m < \infty; T \geq j\}.$$

Therefore by (2.31) - (2.34),

$$P(W \cap V^+) = P(W)P(V^+). \quad \square$$

Clearly $P(V^+) \leq P(V^-)$. But

$$P(V^+) \geq \frac{1}{2d}P(V^+ \mid S_1 = \langle -1 \rangle) = \frac{1}{2d}P(V^-).$$

Therefore by (2.27) and Proposition 2.4.6, as $\lambda \to 1-$,

$$P(V^+) \asymp \begin{cases} (1 - \lambda)^{1/4}, & d = 2, \\ (\ln(\frac{1}{1-\lambda}))^{-1/2}, & d = 3. \end{cases}$$

It then follows from Theorem 2.4.3 that

$$P\{\tau_{U^+} > n\} \asymp \begin{cases} n^{-1/4}, & d = 2, \\ (\ln n)^{-1/2}, & d = 3. \end{cases} \tag{2.35}$$

Consider $P\{\tau_{U^+} > \xi_n\}$. Since a random walk takes about n^2 steps to reach ∂C_n, we would like to conclude that

$$P\{\tau_{U^+} > \xi_n\} \asymp \begin{cases} n^{-1/2}, & d = 2, \\ (\ln n)^{-1/2}, & d = 3. \end{cases} \tag{2.36}$$

However, this does not follow from (2.35) alone. For example if we let $A = A_n = U^+ \cup \partial C_{n-1}$ one can check that

$$P\{\tau_A > n^2\} \asymp \begin{cases} n^{-1/2}, & d = 2 \\ (\ln n)^{-1/2}, & d = 3, \end{cases}$$

But clearly $P\{\tau_A > \xi_n\} = 0$. However, this example is an exception, and as long as the set is not too bad, we can use the intuitive reasoning.

Proposition 2.4.7 *Suppose $0 \in A \subset Z^d$ and*

$$P\{\tau_A > n^2\} \leq c_2 n^{-\alpha} L(n),$$

where $\alpha \geq 0$, and L is a slowly varying function. Suppose for every $m \geq n$,

$$P\{\tau_A > m^2\} \geq c_1 m^{-\alpha} L(m).$$

Then there exists a $K = K(c_1, c_2, \alpha, L) > 0$ such that

$$P\{\tau_A > \xi_n\} \geq K n^{-\alpha} L(n).$$

Proof. By the central limit theorem,

$$\sup_{3 \leq m < \infty} P\{\xi_{2m} > m^2\} \leq \sup_{3 \leq m < \infty} P\{|S(m^2)| \leq 2m\}$$
$$= \rho < 1.$$

Therefore for any $x \in C_m$ $(m \geq 3)$,

$$P^x\{\xi_m > m^2\} \leq P\{\xi_{2m} > m^2\} \leq \rho.$$

By the Markov property, if J is any positive integer, $n \geq 3$,

$$P\{\tau_A > Jn^2, \xi_n > Jn^2\} \leq P\{\tau_A > n^2\}\rho^{J-1}$$
$$\leq c_2 n^{-\alpha} L(n)\rho^{J-1}.$$

Therefore,

$$P\{\tau_A > \xi_n\} \geq P\{\xi_n \leq Jn^2, \tau_A > Jn^2\}$$
$$= P\{\tau_A > Jn^2\} - P\{\tau_A > Jn^2, \xi_n > Jn^2\}$$
$$\geq c_1(\sqrt{J}n)^{-\alpha}L(\sqrt{J}n) - c_2\rho^{J-1}n^{-\alpha}L(n).$$

Choose J so that $c_1(\sqrt{J})^{-\alpha} - c_2\rho^{J-1} > 0$ and find $b < 1$ such that $bc_1(\sqrt{J})^{-\alpha} - c_2\rho^{J-1} > 0$. For n sufficiently large, $L(\sqrt{J}n) \geq bL(n)$ and hence

$$P\{\tau_A > \xi_n\} \geq \{bc_1(\sqrt{J})^{-\alpha} - c_2\rho^{J-1}\}n^{-\alpha}L(n).$$

This proves the result for n sufficiently large. For n small, one has only a finite number of cases and can handle them easily, changing the constant if necessary. $\square$

From Proposition 2.4.7 and (2.35) we have

$$P\{\xi_n < \tau_{U^+}\} \geq \begin{cases} cn^{-1/2}, & d = 2 \\ c(\ln n)^{-1/2}, & d = 3. \end{cases} \tag{2.37}$$

What we wish to conclude is that

$$P\{\xi_{2n} < \tau_{U_n^+}\} \asymp \begin{cases} cn^{-1/2}, & d = 2, \\ c(\ln n)^{-1/2}, & d = 3. \end{cases}$$

Let $\partial C_n = B^+ \cup B^- \cup B^0$, where

$$B^+ = \{(z_1, \ldots, z_d) \in \partial C_n : z_1 > 0\},$$

and similarly for B^-, B^0. By symmetry,

$$P\{S(\xi_n) \in B^+\} = P\{S(\xi_n) \in B^-\}.$$

Suppose $x \in U_n^+$, and suppose a random walk starting at x hits B^- before hitting $B^+ \cup B^0$. Then it is easy to see that the random walk path starting at x which takes the negative of every step the first path takes hits B^+ before hitting $B^- \cup B^0$. Therefore for every $x \in U_n^+$,

$$P^x\{S(\xi_n) \in B^+\} \geq P^x\{S(\xi_n) \in B^-\}.$$

Since this is true for every $x \in U_n^+$,

$$P\{S(\xi_n) \in B^+ \mid \tau_{U_n^+} < \xi_n\} \geq P\{S(\xi_n) \in B^- \mid \tau_{U_n^+} < \xi_n\},$$

and hence

$$P\{S(\xi_n) \in B^- \cup B^0 \mid \tau_{U_n^+} > \xi_n\} \geq \frac{1}{2}. \tag{2.38}$$

By the Harnack principle (Theorem 1.7.6), for $x \in B^- \cup B^0$,

$$P^x\{\xi_{2n} < \tau_{U_n^+}\} \geq c;$$

Therefore, by the strong Markov property,

$$P\{\xi_{2n} < \tau_{U_n^+}\} \geq \begin{cases} cn^{-1/2}, & d = 2, \\ c(\ln n)^{-1/2}, & d = 3. \end{cases} \tag{2.39}$$

To get the inequality in the other direction we need an easy lemma.

Lemma 2.4.8 *Let* $S_n = (S_n^1, \ldots, S_n^d)$ *be a simple random walk in* Z^d. *Then for every* $a > 0$, *there exists a* $c_a > 0$ *such that for* $n \geq 2$,

$$P\{|S_j^i| < n; i = 1, \ldots, d; j = 0, \ldots, an^2\} \geq c_a.$$

Proof. We will prove the result for $d = 1$; for $d > 1$ one can consider each component separately. We may also assume without loss of generality that 4 divides n. Let $\tau = \inf\{j \geq 1 : |S_j| \geq n\}$. By (1.21), if $|x| < \frac{n}{2}$, $E^x(\tau) \geq \frac{n^2}{2}$. Let $\eta = \inf\{j \geq 1 : |S_j| \geq \frac{n}{2}\}$. Then $E(\eta) = \frac{n^2}{4}$. Also by (1.21),

$$\frac{n^2}{4} \leq E(\eta) = E(\eta \mid \eta > \frac{n^2}{8})P\{\eta > \frac{n^2}{8}\}$$

$$+ E(\eta \mid \eta \leq \frac{n^2}{8})P\{\eta \leq \frac{n^2}{8}\}$$

$$\leq (\frac{n^2}{8} + \frac{n^2}{4})P\{\eta > \frac{n^2}{8}\} + \frac{n^2}{8}P\{\eta \leq \frac{n^2}{8}\},$$

and hence for n sufficiently large,

$$P\{\eta > \frac{n^2}{8}\} \geq \frac{1}{4}.$$

For any $|x| < \frac{n}{2}$, the above implies

$$P^x\{|S_j - x| < \frac{n}{2}, j = 0, 1, \ldots, \frac{n^2}{8}\} \geq \frac{1}{4}.$$

If this event occurs, then the path stays in C_n. Also if we consider any such path, either it ends in $C_{n/2}$ or the negative of the path ends in $C_{n/2}$. Therefore, since $\frac{n^2}{8}$ is an integer, if $|x| < \frac{n}{2}$,

$$P^x\{\tau > \frac{n^2}{8}, S(\frac{n^2}{8}) \in C_{n/2}\} \geq \frac{1}{8},$$

and therefore by the Markov property if $\frac{k}{8} \leq a \leq \frac{k+1}{8}$,

$$P\{\tau > an^2\} \geq (\frac{1}{8})^{k+1}. \quad \square$$

Corollary 2.4.9 *If $A \subset C_n$, then*

$$P\{\tau_A > \xi_{2n}\} \leq cP\{\tau_A > n^2\}.$$

Proof. By Lemma 2.4.8, for $x \in \partial C_{2n}$,

$$P^x\{\tau_A > n^2\} \geq c,$$

and hence

$$P\{\tau_A > n^2 \mid \tau_A > \xi_{2n}\} > c. \quad \square$$

If we return to the line segment, another application of Lemma 2.4.8 gives

$$P\{\tau_{U_n^+} > n^2 \mid \xi_n < \tau_{U_n^+}, S(\xi_n) \in B^- \cup B^0\} \geq c,$$

and therefore, by (2.38),

$$P\{\tau_{U_n^+} > n^2\} \geq cP\{\xi_n < \tau_{U_n^+}\}.$$

If we combine this with (2.39), we get

$$P\{\xi_{2n} < \tau_{U_n^+}\} \asymp \begin{cases} n^{-1/2}, & d = 2, \\ (\ln n)^{-1/2}, & d = 3. \end{cases} \tag{2.40}$$

Finally by Exercise 2.1.4, this gives

$$H_{U_n^+}(0) \asymp \begin{cases} n^{-1/2}, & d = 2, \\ (\ln n)^{1/2}n^{-1}, & d = 3. \end{cases} \tag{2.41}$$

In the next section we will need an estimate of the harmonic measure of U_n^+ at points near 0 for $d = 2$.

Proposition 2.4.10 *For $0 < m < n$, $d = 2$,*

$$H_{U_n^+}(\langle m \rangle) \leq cm^{-1/2}n^{-1/2}.$$

Proof. By (2.30), it suffices to prove the result for $m \leq \frac{n}{16}$. Therefore assume $m \leq \frac{n}{16}$ and let $f(x) = P^x\{\tau_{U_n^+} > \xi_{2n}\}$. We will prove that for $x \in \partial C_{2m}$,

$$f(x) \leq cm^{1/2}n^{-1/2}. \tag{2.42}$$

From (2.42) we can conclude that

$$
\begin{aligned}
f(\langle m \rangle) &\leq P^{\langle m \rangle}\{\xi_{2m} < \tau_{U_n^+}\} \sup_{x \in \partial C_{2m}} f(x) \\
&\leq (cm^{-1})(cm^{1/2}n^{-1/2}) \\
&\leq cm^{-1/2}n^{-1/2}.
\end{aligned}
$$

As above, write $\partial C_{2m} = B^+ \cup B^- \cup B^0$. For $x, y \in B = B^- \cup B^0$, the Harnack principle gives

$$f(x) \leq cf(y). \tag{2.43}$$

Also (2.38) and (2.37) give

$$P\{\xi_{2m} < \tau_{U_n^+}; S(\xi_{2m}) \in B\} \geq cm^{-1/2}. \tag{2.44}$$

Since

$$P\{\xi_{2n} < \tau_{U_n^+}\} \leq cn^{-1/2},$$

(2.43) and (2.44) imply

$$f(x) \leq cm^{1/2}n^{-1/2}, \quad x \in B.$$

Similarly if $W = \{(z_1, z_2) : -8m \leq z_1 \leq -m, |z_2| \leq 4m\}$, the Harnack principle gives

$$f(x) \leq cm^{1/2}n^{-1/2}, \quad x \in W.$$

If we consider the set $K = \{\langle j \rangle : 4m \leq j < 4m + \frac{n}{2}\}$ and the ball $\tilde{C} = \{x + \langle 4m \rangle : x \in C_n\}$, we get by above, if $x \in \partial C_{2m}$,

$$
\begin{aligned}
f(x) &\leq P^x\{\tau_K > \tau_{\tilde{C}}\} \\
&\leq cm^{1/2}n^{-1/2}.
\end{aligned}
$$

Therefore, (2.42) holds and the proposition is proved. □

2.5   Upper Bounds for Harmonic Measure

Given a finite set A, $x \in A$, how large can $H_A(x)$ be? If A consists of two points, then by symmetry $H_A(x) = \frac{1}{2}$, and hence by (2.11) if A has at least two points

$$H_A(x) \le \frac{1}{2}.$$

For $d = 2$, we cannot improve on the above estimate if we only know the cardinality of A. To see this, we first note that if $A \subset C_m$ and

$$\sigma = \sigma_m = \sup\{j \le \xi_m : S_j \in A\},$$

then (see the proof of Proposition 2.4.1)

$$P\{S_\sigma = x\} = G_{C_m}(0, x) P^x\{\tau_A > \xi_m\}.$$

Hence by Proposition 1.6.7 and Theorem 2.1.3 ,

$$\lim_{m \to \infty} P\{S_\sigma = x\} = H_A(x). \tag{2.45}$$

Now let $A = C_n \cup \{x\}$ where $x = x_n = (2^n, 0)$. By Lemma 2.3.1 and Theorem 1.6.2, if $y \in C_n$,

$$\lim_{m \to \infty} (\ln m) P^y\{\tau_x > \xi_m\} = n \ln 2 + o(\ln n),$$
$$\lim_{m \to \infty} (\ln m) P^x\{\xi_n > \xi_m\} = n \ln 2 + O(\ln n).$$

By considering the successive times that the random walk is in C_n, then x, then C_n, then x, etc., one can see that this implies

$$\lim_{m \to \infty} P\{S_\sigma = x\} = \frac{1}{2} - O(\frac{\ln n}{n}),$$

or by (2.45),

$$H_A(x) = \frac{1}{2} - O(\frac{\ln n}{n}).$$

If $d \ge 3$, we can get a nontrivial estimate of the harmonic measure of a point in terms of $|A|$, where $|\cdot|$ denotes cardinality.

Proposition 2.5.1 If $A \subset Z^d, x \in A$, then

$$H_A(x) \le c|A|^{(2-d)/d}.$$

Proof. Let
$$v(n) = \sup_{|A|=n} \sum_{x \in A} G(x).$$

Then by Theorem 1.5.4, for n sufficiently large if $m = 2n^{1/d}$,

$$\begin{aligned}
v(n) &\le \sum_{x \in C_m} G(x) \\
&\le c \sum_{x \in C_m} (|x|^{2-d} \wedge 1) \\
&\le cn^{2/d}.
\end{aligned}$$

Now assume $|A| = n$ and let $v = v(n)$. By Proposition 2.4.1(d), for $x \in A$,

$$\sum_{y \in A} G(x,y)(\mathrm{Es}_A(y) - v^{-1}) = 1 - v^{-1} \sum_{y \in A} G(y-x) \ge 0.$$

Therefore, again using Proposition 2.4.1(d),

$$\begin{aligned}
0 &\le \sum_{x \in A} \mathrm{Es}_A(x) \sum_{y \in A} G(x,y)(\mathrm{Es}_A(y) - v^{-1}) \\
&= \sum_{y \in A} (\mathrm{Es}_A(y) - v^{-1}) \sum_{x \in A} \mathrm{Es}_A(x) G(x,y) \\
&= \sum_{y \in A} (\mathrm{Es}_A(y) - v^{-1}) \sum_{x \in A} G(y,x) \mathrm{Es}_A(x) \\
&= \sum_{y \in A} (\mathrm{Es}_A(y) - v^{-1}) \\
&= \mathrm{cap}(A) - nv^{-1}.
\end{aligned}$$

Therefore
$$\mathrm{cap}(A) \ge n[v(n)]^{-1} \ge cn^{(d-2)/d},$$

and by (2.13),
$$H_A(x) \le cn^{(2-d)/d}. \quad \square$$

If $A_x = C_n \cup \{x\}$, then $|A_x| \asymp n^d$. By (2.16) and Proposition 2.2.1, $\mathrm{cap}(A_x) \asymp n^{d-2}$. But by transience,

$$\lim_{|x| \to \infty} \mathrm{Es}_{A_x}(x) = \mathrm{Es}_x(x) > 0,$$

Therefore, for $|x|$ sufficiently large,

$$H_A(x) \asymp n^{2-d} \asymp |A|^{(2-d)/d},$$

so the bound in Proposition 2.5.1 cannot be improved if one only knows the cardinality of A.

We now consider how large the harmonic measure can be for points in a connected set of a certain radius. Let $\mathcal{A}_n$ be the set of all connected subsets of Z^d of radius n which contain 0. Let $\mathcal{B}_n$ be the set of subsets of Z^d of radius n which for each $j = 1, 2, \ldots, n$, contain exactly one point x with $j - 1 \leq |x| < j$. As we have noted previously, if $A \in \mathcal{A}_n$, there exists a (perhaps disconnected) $B \subset A$ with $B \in \mathcal{B}_n$. In the previous section we considered the line segment $A = U_n^+$ and showed

$$H_A(0) \leq \begin{cases} cn^{-1/2}, & d = 2, \\ c(\ln n)^{1/2}n^{-1}, & d = 3, \\ cn^{-1}, & d \geq 4. \end{cases}$$

One might guess that the line segment is as sparse a connected set as one could have, and that the endpoint of the segment has the largest possible harmonic measure for a connected set. The remainder of this section will be devoted to proving that this is the case, at least up to a multiplicative constant. This theorem is a discrete analogue of the Beurling projection theorem [1].

Theorem 2.5.2 *If $0 \in A \subset Z^d$ is a connected set of radius n then,*

$$H_A(0) \leq \begin{cases} cn^{-1/2}, & d = 2, \\ c(\ln n)^{1/2}n^{-1}, & d = 3, \\ cn^{-1}, & d \geq 4. \end{cases}$$

The following lemma was proved in the proof of Lemma 2.3.5.

Lemma 2.5.3 *If $d = 2$, $B \in \mathcal{B}_n$ and $x \in \partial C_n$,*

$$P^x\{\tau_B < \xi_{2n}\} \geq c.$$

The next lemma gives a similar result for $d \geq 3$.

Lemma 2.5.4 *If $d \geq 3$, $B \in \mathcal{B}_n$, then*

$$\text{cap}(B) \geq \begin{cases} cn(\ln n)^{-1}, & d = 3, \\ cn, & d \geq 4. \end{cases}$$

Proof. Let Y be the number of visits to B, i.e.,

$$Y = \sum_{x \in B} \sum_{j=0}^{\infty} I\{S_j = x\}.$$

By Theorem 1.5.4, if $y \in \partial C_{2n}$,

$$E^y(Y) \geq c \sum_{x \in B} n^{2-d} = cn^{3-d} \tag{2.46}$$

If $y \in C_n$, there exist at most $2j$ points of B within distance $j - 1$ of y and hence by Theorem 1.5.4,

$$E^y(Y) \leq c \sum_{j=1}^{n} j^{2-d} \leq \begin{cases} c(\ln n), & d = 3, \\ c, & d \geq 4. \end{cases} \tag{2.47}$$

If $y \in \partial C_{2n}$,

$$E^y(Y) = P^y\{\tau_B < \infty\}E^y(Y \mid \tau_B < \infty).$$

Therefore, by (2.46) and (2.47),

$$P^y\{\tau_B < \infty\} \geq \begin{cases} c(\ln n)^{-1}, & d = 3, \\ cn^{3-d}, & d \geq 4. \end{cases}$$

But by Proposition 2.2.2,

$$\mathrm{cap}(B) \asymp n^{d-2} P^y\{\tau_B < \infty\},$$

which gives the lemma. $\square$

Proof of Theorem 2.5.2. By (2.11), it suffices to prove the result for $B \in \mathcal{B}_n$. If $d \geq 4$, the theorem follows immediately from (2.13) and Lemma 2.5.4. Consider $d = 2$. Let $m = n^3$. Then by Theorem 1.6.6, Proposition 1.6.7, and Exercise 1.6.8,

$$\begin{aligned}
(\frac{6}{\pi} \ln n)^{-1} &\sim P\{\tau_0 > \xi_m\} \\
&= \sum_{y \in \partial C_m} P\{S(\tau_0 \wedge \xi_m) = y\} \\
&= \sum_{y \in \partial C_m} P^y\{\tau_0 < \xi_m\} \\
&= \sum_{y \in \partial C_m} P^y\{\xi_n < \xi_m\}P^y\{\tau_0 < \xi_m \mid \xi_n < \xi_m\} \\
&\sim \frac{2}{3} \sum_{y \in \partial C_m} P^y\{\xi_n < \xi_m\}.
\end{aligned}$$

Therefore by Lemma 2.5.3,

$$\sum_{x \in B} P^x\{\tau_B > \xi_m\} = \sum_{y \in \partial C_m} P^y\{\tau_B < \xi_m\}$$

$$\asymp \sum_{y \in \partial C_m} P^y\{\xi_n < \xi_m\}$$

$$\asymp (\ln n)^{-1}. \tag{2.48}$$

To prove the theorem it then suffices by Exercise 2.1.4 to prove

$$P\{\tau_B > \xi_m\} \le c(\ln n)^{-1} n^{-1/2}.$$

Let $B = \{y_1, \ldots, y_n\}$, where $j - 1 \le |y_j| < j$. Let $U = U_n^+ = \{z_1, \ldots, z_n\}$ where $z_j = (j - 1, 0)$. By Proposition 2.4.10,

$$H_U(z_j) \le cn^{-1/2} j^{-1/2},$$

and hence by (2.48) and Theorem 2.1.3,

$$P^{z_j}\{\xi_m < \tau_U\} \le c(\ln n)^{-1} n^{-1/2} j^{-1/2}. \tag{2.49}$$

For notational ease let

$$g(x, y) = G_{C_m}(x, y),$$

$$e(x, A) = P^x\{\tau_A > \xi_m\},$$

$$h(x, A) = P^x\{S_j \in A \text{ for some } j = 0, 1, 2, ..., \xi_m\}.$$

We may assume without loss of generality that $u = (-1, 0) \notin B$. We will show

$$e(u, B) \le c(\ln n)^{-1} n^{-1/2},$$

which will clearly imply

$$e(0, B) \le c(\ln n)^{-1} n^{-1/2}.$$

By (2.49), $e(u, U) \le c(\ln n)^{-1} n^{-1/2}$. Therefore,

$$e(u, B) \le c(\ln n)^{-1} n^{-1/2} + e(u, B) - e(u, U).$$

By Proposition 2.4.1(c),

$$\begin{aligned}
e(u, B) - e(u, U) &= h(u, U) - h(u, B) \\
&= \sum_{j=1}^n g(u, z_j) e(z_j, U) - \sum_{j=1}^n g(u, y_j) e(y_j, B) \\
&= \sum_{j=1}^n [g(u, z_j) - g(u, y_j)] e(z_j, U) \\
&\quad + \sum_{j=1}^n g(u, y_j)[e(z_j, U) - e(y_j, B)]. \quad (2.50)
\end{aligned}$$

By Proposition 1.6.7, if $x, y \in C_n$,

$$g(x, y) = \frac{2}{\pi}[3 \ln n - \ln |x - y| + o(|x - y|^{-3/2})].$$ (2.51)

Since $j - 2 \leq |u - y_j| \leq |u - z_j| = j + 1$,

$$g(u, z_j) - g(u, y_j) \leq cj^{-3/2},$$ (2.52)

and therefore,

$$\sum_{j=1}^{n}[g(u, z_j) - g(u, y_j)]e(z_j, U) \leq c \sum_{j=1}^{n} j^{-3/2}(\ln n)^{-1}n^{-1/2}j^{-1/2}$$

$$\leq c(\ln n)^{-1}n^{-1/2}.$$ (2.53)

For the second term consider the function

$$F(x) = \sum_{j=1}^{n} g(x, y_j)[e(z_j, U) - e(y_j, B)].$$

F is harmonic on $C_m \setminus B$, and $F(x) = 0$ for $x \in \partial C_m$. Therefore, by the maximum principle (Exercise 1.4.7),

$$F(u) \leq 0 \vee \sup_{y \in B} F(y).$$

We will show that

$$F(y) \leq c(\ln n)^{-1/2}n^{-1/2},$$

for each $y \in B$, which then implies that the estimate holds for each $y \in C_m$. By Proposition 2.4.1(c),

$$\sum_{j=1}^{n} g(y_i, y_j)e(y_j, B) = 1 = \sum_{j=1}^{n} g(z_i, z_j)e(z_j, U).$$

Therefore,

$$F(y_i) = \sum_{j=1}^{n}[g(y_i, y_j) - g(z_i, z_j)]e(z_j, U).$$

Note that $|y_i - y_j| \geq |j - i| - 1 = |z_i - z_j| - 1$. Therefore by (2.51),

$$g(y_i, y_j) - g(z_i, z_j) \leq c|j - i + 1|^{-1}.$$

We now use (2.49) to give

$$F(y_i) = \sum_{j=1}^{n} c|j-i+1|^{-1}(\ln n)^{-1} n^{-1/2} j^{-1/2}$$
$$\leq cn^{-1/2}(\ln n)^{-1}.$$

The $d = 3$ case is proved similarly. In this case we let

$$g(x,y) = G(x,y),$$

$$e(x,A) = \text{Es}_A(x),$$

$$h(x,A) = P^x\{S_j \in A \text{ for some } j = 0,1,2,\ldots\}.$$

By (2.13) and Lemma 2.5.4, it suffices to prove that

$$e(u,B) \leq c(\ln n)^{-1/2},$$

where $u = (1,0,0)$. By (2.40),

$$e(z,U) \leq c(\ln n)^{-1/2}, z \in U.$$

Then (2.50) holds again. In this case we have the estimate (Theorem 1.5.4)

$$g(x,y) = a_3|x-y|^{-1} + o(|x-y|^{-2}),$$

so that

$$g(u,z_j) - g(u,y_j) \leq cj^{-2}.$$

Then (2.53) gives

$$\sum_{j=1}^{n} [g(u,z_j) - g(u,y_j)]e(z_j,U) \leq c(\ln n)^{-1/2},$$

and similarly for the second term. □

2.6 Diffusion Limited Aggregation

We will give a brief introduction to a model for dendritic growth, first introduced by Witten and Sander, called *diffusion limited aggregation (DLA)*. In this model, one builds a random cluster of points A_n in Z^d according to the following rule:

- $A_1 = \{0\}$.

- if A_n is given, then for $x \in \partial A_n$,

$$P\{A_{n+1} = A_n \cup \{x\} \mid A_n\} = H_{\partial A_n}(x).$$

In other words, a random walker is sent "from infinity" until it reaches a lattice point which is adjacent to the cluster, at which time the points adds onto the cluster. The above rule defines a Markov chain whose state space is the set of finite connected subsets of Z^d containing 0. Note that A_n always has cardinality n.

Computer simulations of the model show that the clusters formed are relatively sparse and appear to have a noninteger "fractal dimension". The notion of fractal dimension is vague, see [55], but there is a natural intuitve feel for what the "dimension" of a subset of Z^d should be. Suppose $\text{rad}(A_n) = m$. Then A_n contains n points all of which lie in the ball of radius m. For integer k, a k-dimensional subset of C_m will have on the order m^k points. So the "dimension" $\bar{d}$ of the cluster A_n can be defined by

$$n \approx m^{\bar{d}},$$

or

$$\text{rad}(A_n) \approx n^{1/\bar{d}}.$$

The last equation has the advantage that we can make a rigorous mathematical definition: we define the dimension of the DLA cluster $\bar{d}$ in d dimensions to be equal to $\frac{1}{\alpha}$ where

$$\alpha = \limsup_{n \to \infty} \frac{\ln E(\text{rad}(A_n))}{\ln n}.$$

We expect in fact that the limit on the right hand side exists and that almost surely

$$\frac{\ln \text{rad}(A_n)}{\ln n} \longrightarrow \alpha,$$

but proving statements about such quantities is very difficult.

Numerical simulations suggest a value a little less than 1.7 for $\bar{d}$ in two dimensions. There is also a mean-field theory that gives a prediction

$$\bar{d} = \frac{d^2 + 1}{d + 1},$$

which agrees fairly well with simulation. See [70] for a discussion of DLA from a nonrigorous viewpoint.

In this section we will use the results of the previous section to give a rigorous upper bound on α. As the reader can note, the bound is far from the conjectured values.

Theorem 2.6.1 *There exists a $c < \infty$ such that almost surely for n suffi-ciently large*

$$\operatorname{rad}(A_n) \leq \begin{cases} cn^{2/3}, & d = 2, \\ cn^{1/2}(\ln n)^{1/4}, & d = 3, \\ cn^{2/d}, & d \geq 4. \end{cases} \tag{2.54}$$

The proof of Theorem 2.6.1 needs an exponential estimate for geometric random variables. Such results are standard; however, it will be just as easy to prove the result we need as to specialize a more general theorem from the literature.

Lemma 2.6.2 *Suppose $T_1, \ldots, T_n$ are independent geometric random variables with parameter p, i.e., $P\{T_i = j\} = p(1-p)^{j-1}; Y = T_1 + \cdots + T_n; p < \frac{1}{2}$. Then for every $a \geq 2p$,*

$$P\{Y \leq \frac{an}{p}\} \leq (2e^2 a)^n.$$

Proof: The moment generating function of Y is

$$E(e^{tY}) = [pe^t]^n [1 - e^t(1-p)]^{-n}.$$

By the Chebyshev inequality, for any $t > 0$,

$$\begin{aligned} P\{Y \leq \frac{an}{p}\} &\leq \exp\{\frac{ant}{p}\} E(e^{-tY}) \\ &= \exp\{\frac{ant}{p}\} p^n [e^t - (1-p)]^{-n}. \end{aligned}$$

Let $t = \ln(\frac{a(1-p)}{a-p})$. Then

$$\begin{aligned} P\{Y \leq \frac{an}{p}\} &\leq [\frac{a(1-p)}{a-p}]^{an/p} p^n (1-p)^{-n} [\frac{p}{a-p}]^{-n} \\ &\leq [1 + \frac{p}{a-p}]^{an/p} 2^n a^n \\ &\leq [1 + \frac{p}{a-p}]^{2(a-p)n/p} (2a)^n \\ &\leq (2e^2 a)^n. \quad \square \end{aligned}$$

Proof of Theorem 2.6.1. Let

$$h(x) = h_d(x) = \begin{cases} x^{2/3}, & d = 2, \\ x^{1/2}(\ln x)^{1/4}, & d = 3, \\ x^{2/d}, & d \geq 4. \end{cases}$$

We will prove that for some $\tilde{c} > 0$, almost surely for all n sufficiently large

$$\lambda_n \geq \tilde{c}h^{-1}(n), \tag{2.55}$$

where

$$\lambda_n = \inf\{j : \mathrm{rad}(A_j) \geq n\}.$$

Then (2.54) follows easily from (2.55). Note that if $n \geq 2$, $h_3^{-1}(n) \leq 2n^2(\ln n)^{-1/2}$. The argument for $d = 2$ and $d = 3$ will be similar using Theorem 2.5.2. For $d \geq 4$, the argument will instead use Proposition 2.5.1. We will write $A_n = \{a_1, \ldots, a_n\}$ where a_j is the jth point added to the cluster.

Assume $d = 2$ or 3. If $x \in A_m$, then there exists a sequence of points $0 = x_1, \ldots, x_k = x$ and indices $1 = j_1 < j_2 < \cdots < j_k \leq m$, such that $|x_i - x_{i-1}| = 1$ and $a_{j_i} = x_i$ (this can easily be proved by induction on m). If $x \in \partial C_{2n}$, then by considering the end of this sequence one can find points $y_1, \ldots, y_k = x$ and times $j_1 < j_2 < \cdots < j_k \leq m$ with $|y_i - y_{i-1}| = 1, a_{j_i} = y_i$, and $y_1 \in \partial C_n$. Clearly $k \geq n$. Fix $\beta > 0$ (to be determined later) and let V_n be the event

$$V_n = \begin{cases} \{\lambda_{2n} \leq \beta n^{3/2}\}, & d = 2, \\ \{\lambda_{2n} \leq \beta n^2(\ln n)^{-1/2}\}, & d = 3. \end{cases}$$

If $[z] = [z_1, \ldots, z_n]$ is any random walk path, let $W_n([z])$ be the event

$$W_n([z]) = \{\exists\, j_1 < j_2 < \cdots < j_n \leq m \text{ such that } a_{j_i} = z_i\},$$

where $m = m_n = \beta n^{3/2}$ if $d = 2$ and $m = \beta n^2(\ln n)^{-1/2}$ if $d = 3$. Let W_n be the union of $W_n([z])$ over all random walk paths $[z]$ with n points and $z_1 \in \partial C_n$. Then by the discussion above, $V_n \subset W_n$.

Fix $[z]$ with $z_1 \in \partial C_n$ and let

$$\tau_i = \text{ the } j \text{ such that } a_j = z_i,$$

$$\sigma_i = \tau_{i+1} - \tau_i.$$

Since A_{j_i} is a connected set of radius at least n, we know by Theorem 2.5.2 that, conditioned on A_{j_i}, the distribution of σ_i is bounded above by that of a geometric random variable with parameter

$$p = p_{n,d} = \begin{cases} c_1 n^{-1/2}, & d = 2, \\ c_1(\ln n)^{1/2}n^{-1}, & d = 3. \end{cases}$$

By Lemma 2.6.2, for n sufficiently large,

$$P\{\tau_{n-1} \leq \beta c_1 n p^{-1}\} \leq (4e^2\beta c_1)^{n-1},$$

and therefore

$$P(W_n([z])) \leq (4e^2 \beta c_1)^{n-1}.$$

The number of random walk paths with n points starting at a point in ∂C_n is bounded above by $cn^{d-1}(2d)^{n-1}$. Therefore,

$$P(W_n) \leq cn^{d-1}(8de^2 \beta c_1)^{n-1},$$

and if we choose β so that $8de^2 \beta c_1 < 1$,

$$\sum_{n=1}^{\infty} P(W_n) < \infty.$$

Hence by the Borel-Cantelli Lemma, for this β,

$$P\{V_n \text{ i.o.}\} \leq P\{W_n \text{ i.o}\} = 0,$$

which gives (2.55) .

For $d \geq 4$, if

$$V_n = \{\lambda_{2n} - \lambda_n \leq \beta n \lambda^{(d-2)/d}\},$$

then by a similar argument to the above, using Proposition 2.5.1 instead of Theorem 2.5.2, we can prove that for some $\beta > 0$,

$$P\{V_n \text{ i.o.}\} = 0,$$

i.e., almost surely for n sufficiently large,

$$\lambda_{2n} - \lambda_n \geq \beta n \lambda_n^{(d-2)/d}.$$

It is then routine to show that this implies that (2.55) holds. $\square$

Kesten [37] has recently improved Theorem 2.6.1 for $d \geq 4$ by showing that

$$\text{rad}(A_n) \leq cn^{2/(d+1)}.$$

Chapter 3

Intersection Probabilities

3.1 Introduction

We start the study of intersection probabilities for random walks. It will be useful to make some notational assumptions which will be used throughout this book for dealing with multiple random walks. Suppose we wish to consider k independent simple random walks $S^1, \ldots, S^k$. Without loss of generality, we will assume that S^i is defined on the probability space (Ω_i, P_i) and that $(\Omega, P) = (\Omega_1 \times \cdots \times \Omega_k, P_1 \times \cdots \times P_k)$. We will use E_i for expectations with respect to P_i; E for expectations with repect to P; ω_i for elements of Ω_i; and $\omega = (\omega_1, \ldots, \omega_k)$ for elements of Ω. We will write $P^{x_1, \ldots, x_k}$ and $E^{x_1, \ldots, x_k}$ to denote probabilities and expectations assuming $S^1(0) = x_1, \ldots, S^k(0) = x_k$. As before, if the $x_1, \ldots, x_k$ are missing then it is assumed that $S^1(0) = \cdots = S^k(0) = 0$. If $\sigma \leq \tau$ are two times, perhaps random, we let

$$S^i[\sigma, \tau] = \{S^i(j) : \sigma \leq j \leq \tau\},$$

$$S^i(\sigma, \tau) = \{S^i(j) : \sigma < j < \tau\},$$

and similarly for $S^i(\sigma, \tau]$ and $S^i[\sigma, \tau)$.

Let S^1, S^2 be independent simple random walks starting at 0 with killing rate $1 - \lambda$ and killing times T^1, T^2. Let

$$f(\lambda) = P\{S^1(0, T^1] \cap S^2(0, T^2] = \emptyset\},$$

be the probability that the paths do not intersect. If we let A be the random set $S^2(0, T^2]$, then we can write

$$f(\lambda) = E_2(P_1\{T^1 < \tau_A\}).$$

In general, it is easy to compute the expected number of intersections of two paths using the local central limit theorem; however, finding the probability of no intersection is difficult.

For comparison, consider two examples where A is not random. First, assume $A = \{0\}$. Then by Proposition 2.4.1(b),

$$P_1\{T^1 < \tau_A\} = [G_\lambda(0)]^{-1}, \qquad (3.1)$$

i.e., the probability of no intersection of A and $S^1(0, T^1]$ is exactly the inverse of the expected number of intersections. However, it is not always the case that this will be true; in fact, as the second example shows, there are cases where the probability of intersection is of a different order of magnitude than the inverse of the expected number of intersections. Let $d = 2$ and $A = U^+ = \{(n, 0) : n \geq 0\}$. Then by (2.26), as $\lambda \to 1$, the expected number of intersections of A and $S^1(0, T^1]$ is asymptotic to $c(1 - \lambda)^{-1/2}$, while it can be shown easily using (2.41) that the probability of no intersection decays like $c(1 - \lambda)^{1/4}$.

In this chapter we will consider two intersection problems for simple random walks which can be considered "easier" because the answer can be guessed by intuitions such as "probability of no intersection = (expected number)$^{-1}$". These problems can be stated as:

1) Let S^1 start at 0 and S^2 start at x where $|x|$ is approximately $\sqrt{n}$. Find

$$P^{0,x}\{S^1[0, n] \cap S^2[0, n] \neq \emptyset\}.$$

2) Let S^1, S^2, S^3 start at 0. Find

$$P\{S^1(0, n] \cap (S^2[0, n] \cup S^3[0, n]) = \emptyset\}.$$

Chapters 4 and 5 will analyze quantities such as $f(\lambda)$ which are "harder" to estimate and whose answer cannot be guessed only by counting the expected number of intersections.

3.2   Preliminaries

Let S^1, S^2 be independent simple random walks starting at 0 in Z^d with killing rate $1 - \lambda$, $\lambda \in (0, 1]$, and let T^1, T^2 be the corresponding killing times. Let

$$g(\lambda) = P\{S^1(i) \neq S^2(j), (0, 0) \prec (i, j) \preceq (T^1, T^2)\}.$$

Here we write $(i_1, i_2) \preceq (j_1, j_2)$ if $i_1 \leq j_1$ and $i_2 \leq j_2$; $(i_1, i_2) \prec (j_1, j_2)$ if $(i_1, i_2) \preceq (j_1, j_2)$ but $(i_1, i_2) \neq (j_1, j_2)$. We let R_λ be the number of intersection times (including $(0,0)$), i.e.,

$$R_\lambda = \sum_{i=0}^{\infty}\sum_{j=0}^{\infty} I\{S^1(i) = S^2(j), (i,j) \preceq (T^1, T^2)\}.$$

As a rule it is much easier to estimate the expected value of random variables such as R_λ than to estimate probabilities such as g.

Proposition 3.2.1 *As* $\lambda \to 1$,

$$E(R_\lambda) = \begin{cases} c(1-\lambda)^{-3/2} + O((1-\lambda)^{-1/2}), & d = 1, \\ c(1-\lambda)^{-1} + O(\ln\frac{1}{1-\lambda}), & d = 2, \\ c(1-\lambda)^{-1/2} + O(1), & d = 3, \\ c\ln\frac{1}{1-\lambda} + O(1), & d = 4, \\ c + O((1-\lambda)^{(d-4)/2}), & d \geq 5. \end{cases}$$

Proof.

$$
\begin{aligned}
E(R_\lambda) &= \sum_{i=0}^{\infty}\sum_{j=0}^{\infty} P\{S^1(i) = S^2(j), (i,j) \preceq (T^1, T^2)\} \\
&= \sum_{i=0}^{\infty}\sum_{j=0}^{\infty} \lambda^{i+j} P\{S^1(i) = S^2(j)\}.
\end{aligned}
$$

But by reversing S^2 we can see that

$$P\{S^1(i) = S^2(j)\} = P\{S^1(i+j) = 0\}.$$

Therefore,

$$
\begin{aligned}
E(R_\lambda) &= \sum_{i=0}^{\infty}\sum_{j=0}^{\infty} \lambda^{i+j} p_{i+j}(0) \\
&= \sum_{j=0}^{\infty} \lambda^j (j+1) p_j(0) \\
&= \sum_{j=0}^{\infty} \lambda^j j p_j(0) + G_\lambda(0).
\end{aligned}
$$

It is easy to show using Theorem 1.2.1 that as $\lambda \to 1$,

$$G_\lambda(0) = \begin{cases} c + O((1-\lambda)^{(d-2)/2}), & d \geq 3, \\ O(\ln\frac{1}{1-\lambda}), & d = 2, \\ O((1-\lambda)^{-1/2}), & d = 1. \end{cases}$$

Therefore it suffices to consider

$$\sum_{j=0}^{\infty} \lambda^j j p_j(0) = \sum_{j=0}^{\infty} \lambda^j j \overline{p}_j(0) + \sum_{j=0}^{\infty} \lambda^j j E_j(0).$$

By (1.6),

$$\sum_{j=0}^{\infty} \lambda^j j E_j(0) \leq c \sum_{j=0}^{\infty} \lambda^j j^{-(d+2)/2}$$

$$= O(1) + O((1-\lambda)^{d/2}).$$

Therefore we only need to estimate

$$\sum_{j=1}^{\infty} \lambda^{2j} (2j) (\frac{d}{4\pi j})^{d/2}.$$

This calculation is left as an exercise. $\square$

Note that $d = 4$ is the "critical dimension" for the problem. If $d \leq 4$, $E(R_\lambda)$ goes to infinity as $\lambda \to 1$, but for $d > 4$, $E(R_\lambda) < \infty$ even for $\lambda = 1$. We would like to show such critical behavior for $g(\lambda)$ as well; however, it is not so easy to estimate $g(\lambda)$ in terms of $E(R_\lambda)$. (The reader should compare this to (3.1) where the probability of no return to the origin is given in terms of the expected number of returns.) We can get an estimate in one direction.

Proposition 3.2.2 *If $\lambda < 1$, or if $\lambda = 1$ and $d \geq 5$,*

$$g(\lambda) \geq [E(R_\lambda)]^{-1}.$$

Proof. If $i, j \geq 0$, we call (i, j) a *-last intersection if

$$S^1(i) = S^2(j); S^1(i_1) \neq S^2(j_1), (i, j) \prec (i_1, j_1) \preceq (T^1, T^2).$$

With probability one, every pair of paths will have at least one *-last intersection although a pair of paths may have more than one. (If $d \geq 5$ and $\lambda = 1$, the existence of a *-last intersection follows from the fact that with probability one the number of intersections in finite.) Therefore

$$1 \leq \sum_{i=0}^{\infty} \sum_{j=0}^{\infty} P\{(i, j) \text{ is a *-last intersection}\}.$$

But,

$P\{(i,j)$ is a *-last intersection$\}$
$$= P\{S^1(i) = S^2(j); (i,j) \preceq (T^1, T^2);$$
$$S^1(i_1) \neq S^2(j_1), (i,j) \prec (i_1, j_1) \preceq (T^1, T^2)\}$$
$$= \lambda^{i+j} p_{i+j}(0) g(\lambda),$$

and the proposition follows by summing. $\square$

It follows immediately that for $d \geq 5$, $g(1) > 0$, i.e.,

$$P\{S^1(i) \neq S^2(j), (0,0) \prec (i,j) \prec (\infty, \infty)\} > 0. \qquad (3.2)$$

We also get a lower bound for $d \leq 4$,

$$g(\lambda) \geq \begin{cases} c[\ln \frac{1}{1-\lambda}]^{-1}, & d = 4, \\ c(1 - \lambda)^{(4-d)/2}, & d < 4. \end{cases} \qquad (3.3)$$

Proposition 3.2.2 is not good enough to conclude that for $d \leq 4$,

$$P\{S^1(i) \neq S^2(j), (0,0) \prec (i,j) \prec (\infty, \infty)\} = 0.$$

This is true and we could prove it now without too much work. However, since this will follow from the results of the next sections we will not bother to prove it here. A more difficult question is deciding how good the bounds in (3.3) are. If one examines the proof of Proposition 3.2.2, one sees that the inequality arises from the fact that a pair of paths can have many *-last intersections. If it were true that most paths had only a few *-last intersections, then one might guess that the RHS of (3.3) would also give an upper bound (up to a multiplicative constant). It turns out, however, that the bound in (3.3) is not sharp in low dimensions. As an example, let us consider the case $d = 1$ which can be done exactly.

Suppose S^1, S^2 are independent one dimensional random walks and

$$V = \{S^1(i) \neq S^2(j), (0,0) \prec (i,j) \preceq (T^1, T^2)\}.$$

Then it is easy to see that

$$V = \{S^1(i) > 0, 0 < i \leq T^1; S^2(j) < 0, 0 < j \leq T^2\}$$
$$\cup \{S^1(i) < 0, 0 < i \leq T^1; S^2(j) > 0, 0 < j \leq T^2\}.$$

But by (3.1),

$$P\{S^1(i) > 0, 0 < i \leq T^1\} = \frac{1}{2} P\{S^1(i) \neq 0, 0 < i \leq T^1\}$$
$$= \frac{1}{2} [G_\lambda(0)]^{-1}$$
$$\sim c(1 - \lambda)^{\frac{1}{2}}.$$

Therefore as $\lambda \to 1$,

$$P\{S^1(i) \neq S^2(j) : (0,0) < (i,j) \leq (T^1, T^2)\} \sim c(1-\lambda).$$

Note that this is not the same power of $(1-\lambda)$ as given in (3.3).

At times we will want to consider walks with a fixed number of steps. Let R_n be the number of intersections up through time n, i.e.,

$$R_n = \sum_{i=0}^{n} \sum_{j=0}^{n} I\{S^1(i) = S^2(j)\}.$$

The analogue to Proposition 3.2.1 can be proved easily.

Proposition 3.2.3 *As $n \to \infty$,*

$$E(R_n) = \begin{cases} cn^{3/2} + O(n^{1/2}), & d = 1, \\ cn + O(\ln n), & d = 2, \\ cn^{1/2} + O(1), & d = 3, \\ c \ln n + O(1), & d = 4, \\ c + O(n^{(4-d)/2}), & d \geq 5. \end{cases}$$

In section 2.4, Tauberian theorems were used to relate quantities with geometric killing times and quantities for fixed step walks. We will use the theorems in that section, but will need one more easy result to handle multiple walks. If $h(n_1, \ldots, n_k)$ is any nonnegative function on $\{0, 1, \ldots\}^k$, and $T^1, \ldots, T^k$ are independent geometric random variables with rate $1-\lambda$, let

$$\Phi(\lambda) = \Phi_h(\lambda) = \sum_{n_1, \ldots, n_k} P\{T^1 = n_1, \ldots, T^k = n_k\} h(n_1, \ldots, n_k).$$

Lemma 3.2.4 *If $h(n_1, \ldots, n_k)$ is nonincreasing in each variable, $b > 0$, and $h_n = h(n, \ldots, n)$, then*

$$(1 - \lambda^b)^k h_b \leq \Phi(\lambda) \leq k(1-\lambda) \sum_{n=0}^{\infty} \lambda^{kn} h_n.$$

Proof: Note that

$$\begin{aligned} P\{T^1 \vee \cdots \vee T^k \leq b\} &= (P\{T^1 \leq b\})^k \\ &\geq (1 - \lambda^b)^k, \end{aligned}$$

and

$$\begin{aligned} P\{T^1 \wedge \cdots \wedge T^k = n\} &\leq \sum_{j=1}^{k} P\{T^j = n; T^1, \ldots, T^k \geq n\} \\ &= k\lambda^{kn}(1-\lambda). \end{aligned}$$

Since h is nonincreasing in each variable,

$$\begin{aligned}
\Phi(\lambda) &\geq \sum_{n_j \leq b} P\{T^1 = n_1, \ldots, T^k = n_k\} h(b, \ldots, b) \\
&= P\{T^1 \vee \cdots \vee T^k \leq b\} h_b \\
&\geq (1 - \lambda^b)^k h_b,
\end{aligned}$$

and

$$\begin{aligned}
\Phi(\lambda) &\leq \sum_{n=0}^{\infty} P\{T^1 \wedge \cdots \wedge T^k = n\} h(n, \ldots, n) \\
&\leq k(1 - \lambda) \sum_{n=0}^{\infty} \lambda^{kn} h_n. \quad \square
\end{aligned}$$

3.3 Long Range Intersections

Let S^1, S^2 be independent simple random walks starting at 0 and x respectively, and as in the last section, let R_n be the number of intersections up through time n,

$$R_n = \sum_{i=0}^{n} \sum_{j=0}^{n} I\{S_i^1 = S_j^2\}.$$

We let $J_n(x)$ be the expected number of intersections,

$$\begin{aligned}
J_n(x) &= E^{0,x}(R_n) \\
&= \sum_{i=0}^{n} \sum_{j=0}^{n} P^{0,x}\{S_i^1 = S_j^2\} \\
&= \sum_{i=0}^{n} \sum_{j=0}^{n} \sum_{y \in Z^d} P\{S_i^1 = y\} P^x\{S_j^2 = y\} \\
&= \sum_{y \in Z^d} G_n(y) G_n(x - y).
\end{aligned}$$

By Proposition 3.2.3,

$$J_n(0) \sim \begin{cases} cn^{(4-d)/2}, & d < 4, \\ c \ln n, & d = 4, \\ c, & d > 4. \end{cases} \tag{3.4}$$

We consider the case where $|x|$ is of order $\sqrt{n}$.

Proposition 3.3.1 *If* $0 < a < b < \infty$, *there exist* $c_1 = c_1(a, b)$ *and* $c_2 = c_2(a, b)$ *such that if* $a\sqrt{n} \leq |x| \leq b\sqrt{n}$,

$$c_1 n^{(4-d)/2} \leq J_n(x) \leq c_2 n^{(4-d)/2}.$$

Proof.

$$
\begin{aligned}
J_n(x) &= \sum_{i=0}^{n} \sum_{j=0}^{n} P^{0,x}\{S_i^1 = S_j^2\} \\
&= \sum_{i=0}^{n} \sum_{j=0}^{n} p_{i+j}(x) \\
&= \sum_{k=0}^{n} (k+1) p_k(x) + \sum_{k=n+1}^{2n} (2n - k + 1) p_k(x).
\end{aligned}
$$

The estimate is then a straightforward application of the local central limit theorem. □

A more difficult quantity to estimate is the probability that the paths intersect, i.e.,

$$\phi_n(x) = P^{0,x}\{S^1[0, n] \cap S^2[0, n] \neq \emptyset\}.$$

Since $\phi_n(x) = P^{0,x}\{R_n > 0\}$, and

$$E(R_n) = P\{R_n > 0\} E\{R_n \mid R_n > 0\},$$

we get

$$\phi_n(x) = P^{0,x}\{R_n > 0\} = J_n(x)[E^{0,x}\{R_n \mid R_n > 0\}]^{-1}.$$

If we could compute $E^{0,x}\{R_n \mid R_n > 0\}$ we would have the answer. Suppose $S^1[0, n] \cap S^2[0, n] \neq \emptyset$. Then the paths intersect at some point. Once they intersect at some point, one might guess that they should have approximately the same number of intersections as two walks starting at the same point (at least up to a multiplicative constant). Hence one might expect that $P^{0,x}\{R_n > 0\}$ is approximately equal to $cJ_n(x)[J_n(0)]^{-1}$. Making such intuitive arguments rigorous is not so easy because there is no natural stopping time along the paths. One cannot talk of the "first" intersection of the two paths because there are two time scales involved. However, in this case, the intuition gives the right answer.

Theorem 3.3.2 *If* $0 < a < b < \infty$, *there exist* $c_1 = c_1(a, b)$ *and* $c_2 = c_2(a, b)$ *such that if* $a\sqrt{n} \leq |x| \leq b\sqrt{n}$,

$$
\left.
\begin{array}{l}
c_1 \\
c_1 (\ln n)^{-1} \\
c_1 n^{(4-d)/2}
\end{array}
\right\}
\leq P^{0,x}\{S^1[0, n] \cap S^2[0, n] \neq \emptyset\} \leq
\left\{
\begin{array}{ll}
c_2, & d < 4, \\
c_2 (\ln n)^{-1}, & d = 4, \\
c_2 n^{(4-d)/2}, & d > 4.
\end{array}
\right.
$$

The upper bound for $d < 4$ is trivial and for $d > 4$ it follows immediately from Proposition 3.3.1 and the inequality

$$P^{0,x}\{R_n > 0\} \le E^{0,x}(R_n).$$

In this section we will prove the lower bound and in the following section we will prove the upper bound for $d = 4$.

Proof of the lower bound: Let $V = V_n = S^1[0,n] \cap S^2[0,n]$ and $Y = Y_n$, the cardinality of V. If X is any nonnegative random variable,

$$E(X) = P\{X > 0\}E\{X \mid X > 0\}.$$

By Jensen's inequality,

$$
\begin{aligned}
E(X^2) &= P\{X > 0\}E\{X^2 \mid X > 0\} \\
&\ge P\{X > 0\}[E\{X \mid X > 0\}]^2 \\
&= [E(X)]^2[P\{X > 0\}]^{-1},
\end{aligned}
$$

i.e.,

$$P\{X > 0\} \ge \frac{[E(X)]^2}{E(X^2)}. \tag{3.5}$$

We will show that

$$
\begin{aligned}
E^{0,x}(Y) &\ge [G_n(0)]^{-2}J_n(x) \\
E^{0,x}(Y^2) &\le c[G_n(0)]^{-4}J_{2n}(x)J_{2n}(0).
\end{aligned}
$$

The lower bound then follows from (3.4), Proposition 3.3.1, and (3.5).

Let $\bar\tau_y = \inf\{j \ge 0 : S_j^1 = y\}$ and

$$H_n(y) = P\{\bar\tau_y \le n\}.$$

By the strong Markov property,

$$G_n(y) \le H_n(y)G_n(0) \le G_{2n}(y). \tag{3.6}$$

Therefore,

$$
\begin{aligned}
E^{0,x}(Y) &= \sum_{y \in Z^d} P^{0,x}\{y \in V\} \\
&= \sum_{y \in Z^d} H_n(y)H_n(x - y) \\
&\ge [G_n(0)]^{-2} \sum_{y \in Z^d} G_n(y)G_n(x - y) = [G_n(0)]^{-2}J_n(x).
\end{aligned}
$$

By expanding the square,

$$E^{0,x}(Y^2) = \sum_{y \in Z^d} \sum_{z \in Z^d} P^{0,x}\{y, z \in V\}.$$

Note that

$$P\{y, z \in S^1[0, n]\} \leq P\{0 \leq \overline{\tau}_y \leq \overline{\tau}_z \leq n\} + P\{0 \leq \overline{\tau}_z \leq \overline{\tau}_y \leq n\}$$
$$\leq H_n(y)H_n(z-y) + H_n(z)H_n(z-y).$$

Similarly,

$$P^x\{y, z \in S^2[0, n]\} \leq H_n(z-y)[H_n(x-y) + H_n(x-z)].$$

Therefore, using (3.6),

$$G_n(0)^4 E^{0,x}(Y^2) \leq \sum_{y \in Z^d} \sum_{z \in Z^d} G_{2n}(z-y)^2[G_{2n}(y) + G_{2n}(z)]$$
$$[G_{2n}(x-y) + G_{2n}(x-z)]$$
$$= 2\sum_{y \in Z^d} \sum_{z \in Z^d} G_{2n}(z-y)^2 G_{2n}(y)G_{2n}(x-y)$$
$$+2\sum_{y \in Z^d} \sum_{z \in Z^d} G_{2n}(z-y)^2 G_{2n}(y)G_{2n}(x-z).$$

$$\sum_{y \in Z^d} \sum_{z \in Z^d} G_{2n}(z-y)^2 G_{2n}(y)G_{2n}(x-y)$$
$$= \sum_{y \in Z^d} \sum_{w \in Z^d} G_{2n}(w)^2 G_{2n}(y)G_{2n}(x-y)$$
$$= J_{2n}(0)J_{2n}(x).$$

$$\sum_{y \in Z^d} \sum_{z \in Z^d} G_{2n}(z-y)^2 G_{2n}(y)G_{2n}(x-z)$$
$$= \sum_{w \in Z^d} \sum_{z \in Z^d} G_{2n}(w)^2 G_{2n}(z-w)G_{2n}(x-z)$$
$$= \sum_{w \in Z^d} G_{2n}(w)^2 J_{2n}(x-w)$$
$$= \sum_{|w| \leq \frac{1}{2}|x|} G_{2n}(w)^2 J_{2n}(x-w) + \sum_{|w| > \frac{1}{2}|x|} G_{2n}(w)^2 J_{2n}(x-w)$$
$$\leq cJ_{2n}(x)J_{2n}(0) + J_{2n}(0) \sum_{|w| > \frac{1}{2}|x|} G_{2n}(w)^2.$$

The last step uses the estimate in Proposition 3.3.1. If $\xi = \inf\{j : |S_j^1| \geq \frac{1}{2}|x|\}$, then

$$
\sum_{|w|>\frac{1}{2}|x|} G_{2n}(w)^2 \leq E\Big(\sum_{i=\xi}^{\xi+2n}\sum_{j=0}^{2n} I\{S_i^1 = S_j^2\}\Big)
$$
$$
= J_{2n}(S_\xi^1) \leq cJ_{2n}(x).
$$

Combining all of these estimates we get

$$
E^{0,x}(Y^2) \leq c[G_n(0)]^{-4} J_{2n}(x) J_{2n}(0),
$$

which completes the proof of the lower bound. □

3.4 Upper Bound in Four Dimensions

We introduce a random variable which will be very useful for $d = 4$. Let

$$
D_n = \sum_{i=0}^n G(S_i^1),
$$

where G is the standard Green's function. Suppose $\tilde{R}_n$ is the number of intersections of two independent random walks starting at the origin, one of length n and the other of infinite length, i.e.,

$$
\tilde{R}_n = \sum_{i=0}^n \sum_{j=0}^\infty I\{S_i^1 = S_j^2\}.
$$

Then D_n is the conditional expectation of $\tilde{R}_n$ given S^1. In particular,

$$
E(D_n) = E(\tilde{R}_n).
$$

Note that

$$
\begin{aligned}
E(\tilde{R}_n) &= \sum_{i=0}^n \sum_{j=0}^\infty p_{i+j}(0) \\
&= \sum_{i=0}^n (i+1)p_i(0) + \sum_{i=n+1}^\infty (n+1)p_i(0) \\
&= (\frac{2}{\pi})^2 \ln n + O(1) \\
&= 2a_4 \ln n + O(1),
\end{aligned}
$$

where a_4 is as defined in Theorem 1.5.4. The reason D_n is very useful in four dimensions is that the variance of D_n grows significantly more slowly than $E(D_n)^2$.

Proposition 3.4.1 *If* $d = 4$, *as* $n \to \infty$,

$$
\begin{aligned}
(a)\ \ E(D_n) &= 2a_4(\ln n) + O(1), \\
(b)\ \ \mathrm{Var}(D_n) &= O(\ln n),
\end{aligned}
$$

and hence

$$
\mathrm{Var}\left(\frac{D_n}{E(D_n)}\right) = O\left(\frac{1}{\ln n}\right).
$$

Before sketching the proof of Proposition 3.4.1, let us motivate why the variance should be small. Assume for a moment that $n = 2^k$ (so that $k = \log_2 n$) and write

$$
D_n = G(0) + \sum_{j=0}^{k-1} Y_j,
$$

where

$$
Y_j = \sum_{i=2^{j-1}+1}^{2^j} G(S_i^1).
$$

Then,

$$
E(Y_j) = \sum_{i=2^{j-1}+1}^{2^j} \sum_{k=0}^{\infty} p_{i+k}(0) \sim \sum_{i=2^{j-1}+1}^{2^j} 2a_4 i^{-1} \sim 2a_4 \ln 2.
$$

One can guess that because of the different length scales involved that the Y_j are asymptotically independent. Hence D_n is the sum of k asymptotically independent random variables with approximately the same mean. If the variances of the Y_j are uniformly bounded, then one could hope that

$$
\begin{aligned}
\mathrm{Var}(D_n) &= \sum_{j=0}^{k-1} \mathrm{Var}(Y_j) + \sum_{j=0}^{k-1}\left(\sum_{i \neq j} \mathrm{Cov}(Y_i, Y_j)\right) \\
&\sim\ ck\ =\ c(\ln n).
\end{aligned}
$$

The proof we sketch below will not use the above intuition, but instead will be a direct calculation.

Proof. We have already proved (a). To prove (b) we will show by direct calculation that

$$
E(D_n^2) = 4a_4^2(\ln n)^2 + O(\ln n).
$$

We will only sketch the main points, allowing the reader to fill in the appropriate details. Let $S_i = S_i^1$. Since

$$E(D_n^2) = \sum_{i=0}^{n} \sum_{j=0}^{n} E(G(S_i)G(S_j)),$$

we need to estimate $E(G(S_i)G(S_j))$. Assume $i < j$. Then by Theorem 1.5.4, $E(G(S_i)G(S_j))$ is approximately equal to

$$a_4^2 E((|S_i| \vee 1)^{-2}(|S_j| \vee 1)^{-2}),$$

which by the central limit theorem is approximately equal to

$$16 a_4^2 i^{-2} E(|X|^{-2}|X+Y|^{-2}), \tag{3.7}$$

where X and Y are independent normal random variables with covariance I and sI respectively, $s = (j - i)/i$. We can compute this expected value using spherical coordinates. We need one fact: since $f(y) = |x - y|^{-2}$ is harmonic in R^4 for $y \neq x$, the average value of f over the ball of radius r about x equals

$$\begin{cases} |x|^{-2}, & r \leq |x|, \\ r^{-2}, & r \geq |x| \end{cases} \tag{3.8}$$

(see, e.g., [66, Theorem 1.9(b)]). Then $E(|X|^{-2}|X+Y|^{-2})$ equals

$$\int_{R^4} \frac{1}{|x|^2} \Big[\int_{R^4} \frac{1}{|x+y|^2} (2\pi s)^{-2} e^{-|y|^2/2s} dy \Big] (2\pi)^{-2} e^{-|x|^2/2} dx.$$

The interior integral in spherical coordinates, using (3.8), equals

$$\int_0^{|x|} \frac{1}{2s^2} r^3 e^{-r^2/2s} |x|^{-2} dr + \int_{|x|}^{\infty} \frac{1}{2s^2} r^3 e^{-r^2/2s} r^{-2} dr$$

$$= \frac{1}{|x|^2} (1 - e^{-|x|^2/2s}).$$

Therefore,

$$E(|X|^{-2}|X+Y|^{-2}) = \int_0^{\infty} \frac{1}{2r} e^{-r^2/2} (1 - e^{-r^2/2s}) dr$$

$$= \int_0^{\infty} \frac{1}{2r} e^{-sr^2/2} (1 - e^{-r^2/2}) dr.$$

To compute this integral for $s > 0$ note that if

$$F(s) = \int_0^{\infty} \frac{1}{2r} e^{-sr^2/2} (1 - e^{-r^2/2}) dr,$$

then

$$F'(s) = -\int_0^\infty \frac{r}{4} e^{-sr^2/2}(1 - e^{-r^2/2})dr.$$

By integrating by parts we get

$$F'(s) = -\frac{1}{4s(s+1)}.$$

Hence if we integrate and note that $F(s) \to 0$ as $s \to \infty$, we get

$$F(s) = \frac{1}{4}\ln(1 + \frac{1}{s}).$$

Returning to (3.7), we get that $E(G(S_i)G(S_j))$ is approximately

$$4a_4^2 i^{-2}\ln(1 + \frac{i}{j-i}),$$

and hence that $E(D_n^2)$ is approximately

$$2\sum_{i=0}^n \sum_{j=i+1}^n 4a_4^2 i^{-2}\ln(1 + \frac{i}{j-i}) \quad \sim \quad 8a_4^2 \int_1^n \int_t^n t^{-2}\ln(1 + \frac{t}{s-t})ds\,dt$$

$$= \quad 8a_4^2 \int_1^n t^{-2} \int_0^{n-t}\ln(1 + \frac{t}{s})ds\,dt.$$

Direct calculation of the integral gives the result.   □

Proof of the upper bound of Theorem 3.3.2. As mentioned before we only need to consider $d = 4$. Let

$$\tilde{R}_n = \sum_{i=0}^\infty \sum_{j=0}^n I\{S_i^1 = S_j^2\}.$$

We will show for some $c = c(a, b)$,

$$P^{0,x}\{\tilde{R}_n > 0\} \le c(\ln n)^{-1},$$

which will give the result. An estimate such as Proposition 3.3.1 gives for some $c = c(a, b)$,

$$E^{0,x}(\tilde{R}_{2n}) \le c.$$

Let $\tau = \tau_n$ be the stopping time

$$\tau = \inf\{i \ge 0 : S_i^1 \in S^2[0, n]\},$$

and define $\sigma = \sigma_n$ by

$$\sigma = \inf\{j : S^1(\tau) = S_j^2\}.$$

We will say that j is *good* if

$$D_{j,n} \doteq \sum_{k=0}^{n} G(S_{j+k}^2 - S_j^2) \geq 4a_4(\ln n),$$

and *bad* otherwise. By the strong Markov property applied to S^1,

$$E^{0,x}(\tilde{R}_{2n} \mid \tau < \infty, \sigma \text{ good }) \geq 4a_4(\ln n).$$

Therefore,

$$
\begin{aligned}
P^{0,x}\{\tau < \infty, \sigma \text{ good }\} &\leq E^{0,x}(\tilde{R}_{2n})[E^{0,x}(\tilde{R}_{2n} \mid \tau < \infty, \sigma \text{ good })]^{-1} \\
&\leq c(a,b)O((\ln n)^{-1}).
\end{aligned}
$$

By Chebyshev's inequality and Proposition 3.4.1,

$$
\begin{aligned}
P\{D_{j,n} \text{ bad }\} &\leq P\{|D_n - E(D_n)| \geq \frac{1}{2}E(D_n)\} \\
&\leq 4\frac{\text{Var}(D_n)}{E(D_n)^2} = O((\ln n)^{-1}).
\end{aligned}
$$

But,

$$
\begin{aligned}
P\{\tau < \infty, \sigma \text{ bad }\} &= \sum_{i=0}^{\infty}\sum_{j=0}^{n} P^{0,x}\{\tau = i, \sigma = j, j \text{ bad }\} \\
&\leq \sum_{i=0}^{\infty}\sum_{j=0}^{n} P\{S_i^1 = S_j^2, j \text{ bad }\}.
\end{aligned}
$$

But the events $\{S_i^1 = S_j^2\}$ and $\{j \text{ bad}\}$ are independent. Therefore,

$$
\begin{aligned}
P\{\tau < \infty, \sigma \text{ bad }\} &\leq \sum_{i=0}^{\infty}\sum_{j=0}^{n} P^{0,x}\{S_i^1 = S_j^2\}P\{j \text{ bad }\} \\
&\leq O((\ln n)^{-1})E(\tilde{R}_n) \\
&\leq c(a,b)O((\ln n)^{-1}).
\end{aligned}
$$

Therefore,

$$
\begin{aligned}
P^{0,x}\{\tilde{R}_{2n} > 0\} &= P^{0,x}\{\tau < \infty, \sigma \text{ good }\} + P^{0,x}\{\tau < \infty, \sigma \text{ bad }\} \\
&\leq c(a,b)O((\ln n)^{-1}). \quad \square
\end{aligned}
$$

It is a consequence of the proof that for every x

$$P^{0,x}\{\tilde{R}_n > 0\} \leq c(\ln n)^{-1} E^{0,x}(\tilde{R}_n). \qquad (3.9)$$

The proof also suggests a method for estimating precisely the asymptotics of the intersection probability of two random walks in four dimensions, i.e.,

$$P^{0,x}\{\tilde{R}_n > 0\}.$$

Let $\tilde{J}_n(x) = E^{0,x}(\tilde{R}_n)$. Let τ and σ be defined as above. Note that

$$S_1[0, \tau) \cap S_2[0, n] = \emptyset,$$

and hence $S_1[0, \infty) \cap S_2[0, n]$ is equal to

$$(S_1[\tau, \infty) \cap S_2[0, \sigma)) \cup (S_1[\tau, \infty) \cap S_2[\sigma, n]).$$

How many intersections of the paths do we have, given that there is at least one intersection? The random walk after time τ, $S_{n+\tau}^1 - S_\tau^1$, is independent of the choice of τ and σ. Assuming σ is a "good" point (which it is with high probability), the expected number of intersections of S_1 with $S_2[\sigma, n]$ should look like that of two walks starting at the origin, i.e. $\tilde{J}_n(0)$ (here we are using the fact that $\ln n \sim \ln an$ for any a). Similarly $S_1[\tau, \infty)$ should be expected to intersect $S_2[0, \sigma)$ about $\tilde{J}_n(0)$ times. Therefore we could conjecture that $E\{\tilde{R}_n \mid \tilde{R}_n \neq 0\} \approx 2\tilde{J}_n(0)$ and hence that $P^{0,x}\{\tilde{R}_n \neq 0\}$ is approximately $\tilde{J}_n(x)/(2\tilde{J}_n(0))$. For $|x|$ of order $\sqrt{n}$ this intuition is correct. We will prove such a result in Theorem 4.3.5.

3.5   Two-Sided Walks

The quantity we are most interested in understanding is

$$f(n) = P\{S^1(0, n] \cap S^2(0, n] = \emptyset\}.$$

It turns out to be easier to estimate a quantity which appears at first to be more complicated. Assume we have three simple random walks S^1, S^2, S^3 starting at the origin. We can combine S^2 and S^3 into a single two-sided random walk

$$W(j) = \begin{cases} S^2(j), & -\infty < j \leq 0, \\ S^3(j), & 0 \leq j < \infty. \end{cases}$$

Let

$$\begin{aligned} F(n) &= P\{S^1(0, n] \cap W[-n, n] = \emptyset\} \\ &= P\{S^1(0, n] \cap (S^2[0, n] \cup S^3[0, n]) = \emptyset\}. \end{aligned}$$

What we will prove in the next few sections is the following.

Theorem 3.5.1 *If $F(n)$ is defined as above, then*

$$F(n) \asymp \begin{cases} n^{(d-4)/2}, & d < 4, \\ (\ln n)^{-1}, & d = 4, \\ c, & d > 4. \end{cases} \tag{3.10}$$

Note that this theorem states that up to a multiplicative constant $F(n)^{-1}$ is equal to the expected number of intersections of the paths. The proof for $d > 4$ can be done easily using the ideas in section 3.2. For $d = 1$, one can show as in section 3.2 that

$$P\{S^1(i) \neq 0, 0 < i \leq n\} \sim cn^{-1/2},$$

so that

$$\begin{aligned} F(n) &= P\{S^1(0,n] \subset [1,\infty), S^2[0,n] \subset (-\infty,0], S^3[0,n] \subset (-\infty,0]\} \\ &\quad + P\{S^1(0,n] \subset (-\infty,-1], S^2[0,n] \subset [0,\infty), S^3[0,n] \subset [0,\infty)\} \\ &\sim cn^{-3/2}. \end{aligned}$$

We will prove Theorem 3.5.1 for $d = 2, 3$ in the next two sections; the $d = 4$ case will be handled in the next chapter.

The lower bound for $d = 2, 3$ will follow from the result of section 3.3. By Theorem 3.3.2, if $\sqrt{n} \leq |x| \leq 2\sqrt{n}$,

$$P^{x,0,0}\{S^1[0,2n] \cap W[-3n,3n] \neq \emptyset\} \geq c.$$

Suppose that with high probability the intersection of the path does not occur until after time n on the first path and the other path is hit only on the set $W[-2n, 2n]$. To be precise, let B be the event:

$$\begin{aligned} S^1[0,2n] \cap W[-3n,3n] &\neq \emptyset, \\ S^1[0,2n] \cap W[-3n,-2n] &= \emptyset, \\ S^1[0,2n] \cap W[2n,3n] &= \emptyset, \\ S^1[0,n] \cap W[-3n,3n] &= \emptyset, \end{aligned}$$

and suppose for $\sqrt{n} \leq |x| \leq 2\sqrt{n}$,

$$P^{x,0,0}(B) \geq c. \tag{3.11}$$

Let

$$\begin{aligned} \tau &= \inf\{i : S_i^1 \in W[-3n,3n]\}, \\ \sigma &= \inf\{j \geq -3n : W(j) = S^1(\tau)\}. \end{aligned}$$

Then by (3.11), if $\sqrt{n} \le |x| \le 2\sqrt{n}$,

$$\sum_{i=n+1}^{2n} \sum_{j=-2n}^{2n} P^{x,0,0}\{\tau = i, \sigma = j\} \ge c,$$

and hence,

$$\sum_{\sqrt{n} \le |x| \le 2\sqrt{n}} \sum_{i=n+1}^{2n} \sum_{j=-2n}^{2n} P^{x,0,0}\{\tau = i, \sigma = j\} \ge cn^{d/2}.$$

By reversing time on S^1 and translating so that S_j^2 is the origin, one can see that for $n + 1 \le i \le 2n$, $-2n \le j \le 2n$,

$$\sum_{x \in Z^d} P^{x,0,0}\{\tau = i, \sigma = j\} \le F(n).$$

Therefore, by summing over i and j,

$$(4n + 1)nF(n) \ge cn^{d/2},$$

or

$$F(n) \ge cn^{(d-4)/2}.$$

It is therefore sufficient in order to prove the lower bound in Theorem 3.5.1 for $d = 2, 3$ to prove the following lemma.

Lemma 3.5.2 *If B is the set defined above, then for $\sqrt{n} \le |x| \le 2\sqrt{n}$,*

$$P^{x,0,0}(B) \ge c.$$

   Proof: For any $0 < r < K < \infty$, $B \supset A_{r,K}$ where $A_{r,K}$ is the event that the following hold:

(a) $|S_i^1 - x| \le \frac{r}{3}\sqrt{n}$, $0 \le i \le n$,
(b) $S^1[n, 2n] \cap W[-n, n] \ne \emptyset$,
(c) $|W(i) - x| \ge r\sqrt{n}$, $0 \le |i| \le n$,
(d) $|S_i^1| \le K\sqrt{n}$, $0 \le i \le 2n$,
(e) $|W(i) - x| \ge \frac{r}{2}\sqrt{n}$, $n \le |i| \le 2n$,
(f) $|S^2(2n)|, |S^3(2n)| \ge 3K\sqrt{n}$,
(g) $|W(i)| \ge 2K\sqrt{n}$, $2n \le |i| \le 3n$.

It therefore suffices to show that there exist r, K such that

$$P^{x,0,0}(A_{r,K}) \ge c.$$

By Lemma 2.4.8, for any $r > 0$ there is a $c_r > 0$ such that

$$P^x\{(a) \text{ holds}\} \geq c_r. \tag{3.12}$$

It is easy to check using the reflection principle (Exercise 1.3.4) that

$$\lim_{K \to \infty} \inf_n P^x\{(d) \text{ holds}\} = 1, \tag{3.13}$$

Estimates from Chapter 1 can be used to show that

$$\lim_{r \to 0} \inf_n P\{(c) \text{ holds}\} = 1. \tag{3.14}$$

For $\frac{1}{2}\sqrt{n} \leq |x| \leq 3\sqrt{n}$, by Theorem 3.3.2,

$$P^{x,0,0}\{S^1[0,n] \cap W[-n,n] \neq \emptyset\} \geq c.$$

Hence by (3.12), (3.13), and (3.14) and the Markov property, there exist r, K such that

$$P^{x,0,0}\{(a) \text{ - } (d) \text{ hold}\} \geq c.$$

Since $|S_n^2 - x| \geq r\sqrt{n}, |S_n^3 - x| \geq r\sqrt{n}$, one can then easily show that

$$P^{x,0,0}\{(e)\text{-}(f) \text{ hold} \mid (a) \text{ - } (d) \text{ hold}\} \geq c,$$

and finally again by Lemma 2.4.8,

$$P\{(g) \text{ holds} \mid (a) \text{ - } (f) \text{ hold}\} \geq c. \quad \square$$

3.6 Upper Bound for Two-Sided Walks

It will be easier to deal with random walks with killing rather than fixed step walks. If S^1, S^2, S^3 are independent simple random walks with killing rate $1 - \lambda$ and killing times T^1, T^2, T^3, let

$$F(\lambda) = P\{S^1(0, T^1] \cap (S^2[0, T^2] \cup S^3[0, T^3]) = \emptyset\}.$$

In this section we will prove for $d = 2, 3$,

$$F(\lambda) \leq c(1 - \lambda)^{(4-d)/2}. \tag{3.15}$$

To derive the upper bound for Theorem 3.5.1 from (3.15), let

$$F(n_1, n_2, n_3) = P\{S^1(0, n_1] \cap (S^2[0, n_2] \cup S^2[0, n_3]) = \emptyset\}.$$

Then F is decreasing in each variable. By Lemma 3.2.4, for each $\lambda > 0$,

$$F(n) \leq (1 - \lambda^n)^{-3}\Phi(\lambda).$$

But in this case $\Phi(\lambda) = F(\lambda)$ so by letting $\lambda = 1 - \frac{1}{n}$, we get Theorem 3.5.1.

As a step towards deriving (3.15) we will prove a generalization of (1.19) where we replace a point with a (stochastically) translation invariant set. A one-sided (random walk) path of length n is a sequence of points $\Gamma = [x_0, \ldots, x_n]$ with $x_0 = 0, |x_i - x_{i-1}| = 1$. A two-sided path of lengths j and k is a sequence of points $[y_{-j}, \ldots, y_k]$ with $y_0 = 0, |y_i - y_{i-1}| = 1$. We consider two two-sided paths to be different if they have different j and k even if they traverse the same points in the same order. If Γ is a two-sided path of lengths j and k and $-j \le i \le k$, we let $\Phi^i \Gamma$ be the two-sided path of lengths $j + i$ and $k - i$ obtained by translation, i.e., $\Phi^i \Gamma = [z_{-j-i}, \ldots, z_{k-i}]$ where $z_m = y_{m+i} - y_i$. Let Λ be the set of all two-sided paths of finite lengths. A measure $\overline{P}$ on Λ is called *translation invariant* if for every one-sided path Γ of length n and every $0 \le j \le k \le n$,

$$\overline{P}(\Phi^j \Gamma) = \overline{P}(\Phi^k \Gamma).$$

One example of a translation invariant probability measure is the measure induced by two-sided random walk with killing rate $1 - \lambda$. If S^2, S^3 are independent simple walks with killing rate $1 - \lambda$ and killing times T^2, T^3, we let

$$\overline{P}([y_{-j}, \ldots, y_k]) = P\{T^2 = j; T^3 = k; W(i) = y_i, -j \le i \le k\}.$$

It is easy to verify that $\overline{P}$ is translation invariant.

Let S^1 be a simple random walk defined on another probability space (Ω_1, P_1) with killing rate $1 - \lambda$ and killing time T^1. If $\Gamma = [y_{-j}, \ldots, y_k] \in \Lambda$, we let

$$I^+(\Gamma) \quad = \quad \begin{cases} 1 & \text{if } y_i \ne 0, 1 \le i \le k, \\ 0 & \text{otherwise,} \end{cases} \tag{3.16}$$

$$G^\lambda(\Gamma) \quad = \quad \sum_{i=-j}^{k} G_\lambda(y_i), \tag{3.17}$$

and if $\tau = \tau_\Gamma = \inf\{i \ge 1 : S^1(i) \in \{y_{-j}, \ldots, y_k\}\}$,

$$e^\lambda(\Gamma) = P_1\{\tau > T^1\}.$$

Theorem 3.6.1 *If $\overline{P}$ is a translation invariant measure on Λ and $\overline{E}$ denotes expectation with respect to $\overline{P}$, then for every $\lambda \in (0,1)$,*

$$\overline{E}(I^+ G^\lambda e^\lambda) = 1.$$

Proof: Let Γ be a one-sided path of length $|\Gamma|$ and let

$$B_\Gamma = \{\Phi^j\Gamma : 0 \le j \le |\Gamma|\}.$$

It suffices to show that for each Γ,

$$\overline{E}(I^+G^\lambda e^\lambda I_{B_\Gamma}) = \overline{P}(B_\Gamma).$$

Let $\Gamma = [x_0, \dots, x_n]$, $\tilde{\Gamma} = \{x_0, \dots, x_n\}$, and $\tau = \tau_{\tilde{\Gamma}}$,

$$\tau = \inf\{i > 0 : S^1(i) \in \tilde{\Gamma}\}.$$

By Proposition 2.4.1(b), for each j,

$$\sum_{x \in \tilde{\Gamma}} G_\lambda(x_j, x)P^x\{\tau > T^1\} = 1.$$

Therefore, since $\overline{P}$ is translation invariant,

$$
\begin{aligned}
\overline{E}(I^+G^\lambda e^\lambda I_{B_\Gamma}) &= \sum_{j=0}^n (n+1)^{-1}\overline{P}(B_\Gamma)[(I^+G^\lambda e^\lambda)(\Phi^j\Gamma)] \\
&= (n+1)^{-1}\overline{P}(B_\Gamma)\sum_{j=0}^n I^+(\Phi^j\Gamma)P^{x_j}\{\tau > T^1\} \\
&\qquad \sum_{k=0}^n G_\lambda(x_j, x_k) \\
&= (n+1)^{-1}\overline{P}(B_\Gamma)\sum_{k=0}^n \sum_{x \in \tilde{\Gamma}} P^x\{\tau > T^1\}G_\lambda(x_k, x) \\
&= \overline{P}(B_\Gamma). \quad \square
\end{aligned}
$$

The above proof gives a stronger result. Suppose B is a translation invariant subset of Λ, i.e., a set such that

$$B = \{\Phi^j\Gamma : \Gamma \in B\}.$$

Then,

$$\overline{E}(I^+G^\lambda e^\lambda I_B) = \overline{P}(B). \tag{3.18}$$

If Γ is a two-sided walk of lengths $-j$ and k, we let

$$\underline{G}^\lambda(\Gamma) = \inf_{-j \le i \le k} G^\lambda(\Phi^i\Gamma).$$

Then for every a, $B_a = \{\underline{G}^\lambda = a\}$ is translation invariant, and hence by (3.18),

$$
\begin{aligned}
\overline{P}\{\underline{G}^\lambda = a\} &= \overline{E}(I^+ G^\lambda e^\lambda I_{B_a}) \\
&\geq a\overline{E}(I^+ e^\lambda I_{B_a}).
\end{aligned}
$$

Therefore,

$$
\overline{E}(I^+ e^\lambda I_{B_a}) \leq \frac{1}{a}\overline{P}\{\underline{G}^\lambda = a\}.
$$

If we sum over all a (note that Λ is countable and hence $\underline{G}^\lambda$ takes on at most a countable number of values), we get

$$
\overline{E}(I^+ e^\lambda) \leq \overline{E}((\underline{G}^\lambda)^{-1}). \tag{3.19}
$$

We apply (3.19) to the case where $\overline{P} = P_{2,3}$ is the measure on two-sided walks given by S^2, S^3 with killing rate $1 - \lambda$ and killing times T^2, T^3. Then if we write

$$
\begin{aligned}
I^+ &= I^+(S^3[0, T^3]), \\
e &= e^\lambda(W[-T^2, T^3]), \\
Y &= (\underline{G}^\lambda)^{-1}(W[-T^2, T^3]),
\end{aligned}
$$

(3.19) becomes

$$
E_{2,3}(eI^+) \leq E_{2,3}(Y). \tag{3.20}
$$

Note that

$$
F(\lambda) = E_{2,3}(e). \tag{3.21}
$$

Proposition 3.6.2 *For $d = 2, 3$,*

$$
E_{2,3}(eI^+) \leq c(1 - \lambda)^{(4-d)/2}.
$$

Proof: We may assume $\lambda > \frac{1}{2}$. By (3.20), it suffices to prove

$$
E_{2,3}(Y) \leq c(1 - \lambda)^{(4-d)/2}.
$$

Clearly,

$$
\begin{aligned}
E_{2,3}(Y) &= \sum_{j=0}^\infty P_{2,3}\{T^2 + T^3 = j\}E_{2,3}\{Y \mid T^2 + T^3 = j\} \\
&= \sum_{j=0}^\infty (j+1)\lambda^j(1 - \lambda)^2 a_j, \tag{3.22}
\end{aligned}
$$

where
$$a_j = E_{2,3}\{Y \mid T^2 + T^3 = j\}.$$

Let $n = n(\lambda) = [(1 - \lambda)^{-1}]$. Let S be another simple random walk in Z^d, and let $\overline{S}[a, b] = \{S_j - S_a : a \le j \le b\}$. Then if $v = (\underline{G}^\lambda)^{-1}$,

$$a_j = E(v(S[0, j])).$$

For positive integer k,

$$
\begin{aligned}
a_{kn} &= E(v(S[0, kn])) \\
&\le E(\sup_{1 \le j \le k} v(\overline{S}[(j-1)n, jn])) \\
&\le \sum_{j=1}^{k} E(v(\overline{S}[(j-1)n, jn])) \\
&= ka_n.
\end{aligned}
$$

Similarly if $(k-1)n < m \le kn$,

$$a_m \le ka_n.$$

Hence,

$$
\begin{aligned}
\sum_{j=n+1}^{\infty} a_j(j+1)\lambda^j &= \sum_{k=1}^{\infty}\sum_{i=1}^{n} a_{kn+i}(kn+i+1)\lambda^{kn+i} \\
&\le \sum_{k=1}^{\infty}\sum_{i=1}^{n}(k+1)a_n(k+2)n\lambda^{kn} \\
&\le n^2 a_n \sum_{k=1}^{\infty}(k+2)^2(\lambda^n)^k \\
&\le cn^2 a_n. \qquad\qquad (3.23)
\end{aligned}
$$

Let
$$Z_j = \inf_{0 \le i \le k \le j} G_\lambda(S_i - S_k).$$

Then,
$$a_j \le (j+1)^{-1} E(Z_j^{-1}),$$

and since Z_j is decreasing, if $0 \le j \le n$,

$$a_j \le (j+1)^{-1} E(Z_n^{-1}). \qquad\qquad (3.24)$$

Therefore,

$$\sum_{j=0}^{n} a_j(j+1)\lambda^j \leq \sum_{j=0}^{n} \lambda^j E(Z_n^{-1})$$

$$\leq c(1-\lambda)^{-1}E(Z_n^{-1}). \qquad (3.25)$$

If we substitute (3.23) - (3.25) into (3.22), we get

$$E_{2,3}(Y) \leq c(1-\lambda)E(Z_n^{-1}).$$

Therefore it suffices to prove

$$E(Z_n^{-1}) \leq c(1-\lambda)^{(2-d)/2}, \qquad (3.26)$$

where $n = [(1-\lambda)^{-1}]$.

For $\lambda > 1/2$,

$$\lambda^{\frac{1}{1-\lambda}} \geq e^{-2}.$$

Also, by the local central limit theorem, if $x \in Z^d$ with $|x|$ sufficiently large,

$$G_{|x|^2/16}(x) \geq c|x|^{2-d}.$$

Therefore, for all such x, if T is the killing time for S,

$$G_\lambda(x) \geq P\{T \geq |x|^2/16\}G_{|x|^2/16}(x)$$

$$\geq c|x|^{2-d}\exp\{\frac{-|x|^2(1-\lambda)}{8}\}. \qquad (3.27)$$

By changing the constant if necessary, we can see that (3.27) holds for all $x \in Z^d$. Therefore if

$$R = R_n = \sup_{0 \leq i < j \leq n} |S_i - S_j|,$$

we have

$$Z_n^{-1} \leq c|R|^{d-2}\exp\{\frac{R^2}{8(n+1)}\}.$$

But by the reflection principle (Exercise 1.3.4),

$$P\{R \geq r\} \leq P\{\sup_{0 \leq i \leq n} |S_i| \geq \frac{r}{2}\}$$

$$\leq 2P\{|S_n| \geq \frac{r}{2}\}.$$

Therefore,

$$E(Z_n^{-1}) \leq cn^{(d-2)/2} \sum_{r=1}^{\infty} |\frac{r}{\sqrt{n}}| \exp\{\frac{r^2}{8(n+1)}\} P\{\frac{r-1}{2} \leq S_n\}$$

$$\leq cn^{(d-2)/2}$$

$$\leq c(1-\lambda)^{(2-d)/2},$$

which gives (3.26). □
   By Proposition 3.6.2,

$$E_{2,3}(eI^+) \leq c(1-\lambda)^{(4-d)/2}.$$

Note that

$$E_3(I^+) \sim \begin{cases} c[\ln \frac{1}{1-\lambda}]^{-1}, & d = 2, \\ c, & d = 3, \end{cases}$$

so that one might expect that a logarithmic term should appear for $d = 2$ in $F(\lambda) = E_{2,3}(e)$. This is not the case, however, because those paths which have many returns to the origin are very unlikely to be avoided by another path. The next proposition will finish the proof of the upper bound (3.15).

Proposition 3.6.3 *For any d,*

$$E_{2,3}(e) \leq (2d)^2 E_{2,3}(eI^+) + (1-\lambda).$$

Proof: Let $\sigma_0 = 0$ and for $i > 0$,

$$\sigma_i = \inf\{k > \sigma_{i-1} : S_k = 0\}.$$

Let R be the number of returns to the origin by S^3,

$$R = \sup\{j : \sigma_j \leq T^3\}.$$

Then if $I_r = I\{R = r\}$,

$$E_{2,3}(e) = \sum_{r=0}^{\infty} E_{2,3}(eI_r).$$

Note that $I_0 = I^+$. Fix a k and consider the event $\{R = r, \sigma_R = k\}$ with indicator function $I_{r,k}$. Then the sets $S^3[0,k]$ and $S^2[0,T^2] \cup S^3[k,T^3]$ are conditionally independent given $\{I_{r,k} = 1\}$. Let L_k be the indicator function of the complement of the event

$$\{u : |u| = 1\} \subset S^3[0,k].$$

Note that if $L_k = 0$, and $k \leq T^3$, then $e = P\{T^1 = 0\} = 1 - \lambda$. Therefore,

$$E_{2,3}(eI_{r,k}) \leq (1 - \lambda)P(I_{r,k}) + E_{2,3}(eL_kI_{r,k}).$$

But by the conditional independence,

$$E_{2,3}(eL_kI_{r,k}) = E_3(L_kI\{\sigma_r = k\})E_{2,3}(eI_0).$$

Since returns to the origin are independent, and between any two returns a random walk must visit at least one point of norm one,

$$E_3(L_kI\{\sigma_r = k\}) \leq (2d)(\frac{2d - 1}{2d})^r P\{\sigma_r = k\}.$$

Therefore,

$$E_{2,3}(eI_{r,k}) \leq (1 - \lambda)P(I_{r,k}) + 2d(\frac{2d - 1}{2d})^r E_{2,3}(eI^+)P\{\sigma_r = k\},$$

and

$$
\begin{aligned}
E_{2,3}(e) &\leq (1 - \lambda) + \sum_{r=0}^{\infty}\sum_{k=0}^{\infty} 2d(\frac{2d - 1}{2d})^r E_{2,3}(eI^+)P\{\sigma_r = k\} \\
&= (1 - \lambda) + \sum_{r=0}^{\infty} 2d(\frac{2d - 1}{2d})^r E_{2,3}(eI^+) \\
&= (1 - \lambda) + (2d)^2 E_{2,3}(eI^+). \quad \square
\end{aligned}
$$

If

$$
\begin{aligned}
\tilde{F}(\lambda) &= P\{S^1(0, T^1] \cap (S^2(0, T^2] \cup S^3(0, T^3]) = \emptyset\} \\
\tilde{F}(n) &= P\{S^1(0, n] \cap (S^2(0, n] \cup S^3(0, n]) = \emptyset\},
\end{aligned}
$$

an argument such as in the above proof can be given to prove the following corollary of (3.15) and Theorem 3.5.1.

Corollary 3.6.4 *If* $\tilde{F}(n)$ *is defined as above then as* $n \to \infty$,

$$
\tilde{F}(n) \asymp \begin{cases} n^{(d-4)/2}, & d < 4, \\ (\ln n)^{-1}, & d = 4, \\ c, & d > 4. \end{cases}
$$

3.7 One-sided Walks

If $S^1, \ldots, S^{k+1}$ are simple random walks, let

$$f(n, k) = P\{S^1(0, n] \cap (S^2(0, n) \cup \cdots \cup S^{k+1}(0, n]) = \emptyset\}.$$

Note that $f(n)$ as defined in Section 3.5 is $f(n, 1)$, and $\tilde{F}(n)$ as defined in the last section is $f(n, 2)$. If we let $Y = Y_n$ be the random variable on Ω_1 defined by

$$Y = P_2\{S^1(0, n] \cap S^2(0, n] = \emptyset\},$$

then it is easy to see by independence that

$$f(n, k) = E_1(Y^k). \tag{3.28}$$

We can define $f(n, k)$ for noninteger k by (3.28). Since $0 \le Y \le 1$, Hölder's inequality implies that for $j < k$,

$$E_1(Y^k) \le E_1(Y^j) \le [E_1(Y^k)]^{j/k}.$$

By Corollary 3.6.4,

$$f(n, 2) \asymp \begin{cases} n^{(d-4)/2}, & d = 2, 3, \\ (\ln n)^{-1}, & d = 4. \end{cases}$$

This gives an immediate estimate for $f(n, k)$.

Corollary 3.7.1 *(a) If $k < 2$,*

$$\left.\begin{array}{c} c_1 n^{(d-4)/2} \\ c_1 (\ln n)^{-1} \end{array}\right\} \le f(n, k) \le \begin{cases} c_2 n^{k(d-4)/4}, & d = 2, 3, \\ c_2 (\ln n)^{-k/2}, & d = 4. \end{cases}$$

(b) if $k > 2$,

$$\left.\begin{array}{c} c_1 n^{k(d-4)/4} \\ c_1 (\ln n)^{-k/2} \end{array}\right\} \le f(n, k) \le \begin{cases} c_2 n^{(d-4)/2}, & d = 2, 3, \\ c_2 (\ln n)^{-1}, & d = 4. \end{cases}$$

In particular, if $d = 2, 3$,

$$c_1 n^{(d-4)/2} \le f(n) \le c_2 n^{(d-4)/4}, \tag{3.29}$$

and if $d = 4$,

$$c_1 (\ln n)^{-1} \le f(n) \le c_2 (\ln n)^{-1/2}. \tag{3.30}$$

Chapter 4

Four Dimensions

4.1 Introduction

The critical dimension for intersections of two random walks, $d = 4$, will be studied in this chapter. The critical dimension is characterized by logarithmic behavior of the interesting quantities. The results in this chapter will be stronger than what we can prove in two and three dimensions—instead of just upper and lower bounds for probabilities we will be able to give asymptotic expressions.

The starting point will be two results from last chapter, Proposition 3.4.1 and Theorem 3.6.1. These will allow us to give the asymptotic probability of a certain event involving intersections of a one-sided walk and a two-sided walk starting at the origin. As a corollary we will prove Theorem 3.5.1 for $d = 4$. In Section 3, we use this result to give the asymptotics for long-range intersections of two walks.

In section 4 we return to the question of two walks starting at the origin. We are able to determine exactly the exponent of the logarithm in the function

$$f(n) = P\{S^1(0, n] \cap S^2(0, n] = \emptyset\}.$$

The basic idea of the proof relies on two ideas: 1) "short-range" intersections and "long-range" intersections are almost independent in four dimensions (we do not expect this to be true for $d < 4$), and 2) the probabilty of "long-range" intersections can be calculated precisely. These two facts allow us essentially to analyze the "derivative" of $f(n)$ and then to determine the large n behavior.

Four dimensions is the critical dimension for two walks. Suppose instead that we consider k walks, $S^1, \ldots, S^k$ in Z^d. If $d \leq 2$ then the paths of the walks have an infinite number of mutual intersections by recurrence. For $k = 3$ we can show that for $d \geq 4$,

$$P\{S^1(0,\infty) \cap S^2(0,\infty) \cap S^3(0,\infty) = \emptyset\} > 0,$$

while for $d = 3$ the intersection is infinite with probability one. The intuition is the following: a random walk is a "two-dimensional" set (since its intersection with a ball of radius R has on the order of R^2 points) and hence has "codimension" $d-2$. To find the codimension of the intersection of a number of sets one generally adds the codimensions of the sets, i.e., the codimension of the intersection should be $k(d-2)$, assuming this is no greater than d. The critical dimension for k walks can be found by setting

$$k(d-2) = d,$$

getting

$$d = \frac{2k}{k-1}.$$

Note that for $k > 3$, the "critical dimension" lies strictly between 2 and 3. One can make sense of this in a number of ways (see, e.g. [23, 61]), but we will not deal with this in this book. In the last section of this chapter we will consider the case $k = 3, d = 3$ and mention a number of results which are analogous to the case $k = 2, d = 4$.

4.2 Two-sided Walks

We will prove Theorem 3.5.1 for $d = 4$. Let S^1, S^2, S^3 be independent simple random walks in Z^4 with killing rate $1 - \lambda$ and killing times T^1, T^2, T^3. As in section 3.6, we define the following random variables on $\Omega_2 \times \Omega_3$:

$$e = e_\lambda = P_1\{S^1(0, T^1] \cap (S^2[0, T^2] \cup S^3[0, T^3]) = \emptyset\},$$

$$I^+ = I_\lambda^+ = \text{ indicator function of } \{0 \notin S^3(0, T^3]\},$$

$$G = G^\lambda = \sum_{j=0}^{T^2} G_\lambda(S_j^2) + \sum_{k=1}^{T^3} G_\lambda(S_k^3).$$

If $F(\lambda) = P\{S^1(0, T^1] \cap (S^2[0, T^2] \cup S^3[0, T^3]) = \emptyset\}$, then

$$F(\lambda) = E_{2,3}(e).$$

By Theorem 3.6.1,

$$E_{2,3}(eI^+G) = 1. \tag{4.1}$$

In two and three dimensions, G is a nontrivial random variable in the sense that $(EG)^{-1}G$ does not approach a constant random variable as $\lambda \to 1-$. For this reason, we cannot take the G out of the expectation in (4.1) without giving up a multiplicative factor. For $d = 4$, however, $(EG)^{-1}G$ does approach 1 in probability, so it will be relatively easy to pull it out of the expectation. Most of the work in showing that $(EG)^{-1}G \to 1$ was done in Proposition 3.4.1. Here we state without proof the analogous result for killed random walks. It can proved either by following the proof of Proposition 3.4.1 or by using Theorems 2.4.2 and 2.4.3 on Proposition 3.4.1.

Lemma 4.2.1 *As $\lambda \to 1-$, if $E = E_{2,3}$,*

$$
\begin{aligned}
\text{(a)} \quad E(G) &= -4a_4 \ln(1 - \lambda) + O(1), \\
\text{(b)} \quad \text{Var}(G) &= O(-\ln(1 - \lambda)),
\end{aligned}
$$

and therefore

$$
\text{Var}(\frac{G}{EG}) = O(-[\ln(1 - \lambda)]^{-1}).
$$

With this lemma we would now like to say

$$
E_{2,3}(eI^+) \sim [E_{2,3}(G)]^{-1},
$$

and the next theorem shows that this is in fact the case.

Theorem 4.2.2 *As $\lambda \to 1-$,*

$$
E_{2,3}(eI^+) \sim [-4a_4 \ln(1 - \lambda)]^{-1}.
$$

Proof. For any $\epsilon > 0$, let

$$
A = A_{\epsilon,\lambda} = \{|G + 4a_4 \ln(1 - \lambda)| \geq -4a_4\epsilon \ln(1 - \lambda)\},
$$

and let $Y = Y_{\epsilon,\lambda}$ be the indicator function of A. By Lemma 4.2.1 and Chebyshev's inequality, there exists a $c < \infty$ such that

$$
P_{2,3}(A) \leq \frac{c}{\epsilon^2[-\ln(1 - \lambda)]}. \tag{4.2}
$$

We write

$$
E_{2,3}(eI^+) = E_{2,3}(eI^+Y) + E_{2,3}(eI^+(1 - Y)). \tag{4.3}
$$

For an upper bound we use (4.1) to get

$$
\begin{aligned}
E_{2,3}(eI^+) &\leq E_{2,3}(eI^+Y) + [-(1 - \epsilon)4a_4 \ln(1 - \lambda)]^{-1}E_{2,3}(eI^+G) \\
&= E_{2,3}(eI^+Y) + [-(1 - \epsilon)4a_4 \ln(1 - \lambda)]^{-1}. \tag{4.4}
\end{aligned}
$$

Since $eI^+ \leq 1$,

$$E_{2,3}(eI^+Y) \leq E_{2,3}(Y) \leq -\frac{c}{\epsilon^2 \ln(1 - \lambda)},$$

and hence for some $k = k_\epsilon$,

$$E_{2,3}(eI^+) \leq -k[\ln(1 - \lambda)]^{-1}. \tag{4.5}$$

We now show that as $\lambda \to 1-$,

$$E_{2,3}(eI^+Y) = o_\epsilon([-\ln(1 - \lambda)]^{-9/8}). \tag{4.6}$$

Suppose not, i.e., that for some $\epsilon > 0$, some sequence $\lambda_n \to 1$ and some $\alpha = \alpha_\epsilon > 0$,

$$E_{2,3}(eI^+Y) \geq \alpha[-\ln(1 - \lambda_n)]^{-9/8}. \tag{4.7}$$

For ease and without loss of generality we will assume that Ω_i is actually the set of finite random walk paths with the appropriate measure and that $\omega_i(n) = S^i(n, \omega_i)$. We write $\omega_i \cap \omega_j = \emptyset$ if $\omega_i(0, T^i] \cap \omega_j(0, T^j] = \emptyset$. We also let $I(\omega_i)$ be the indicator function of the set $\{0 \notin \omega_i(0, T^i]\}$. For any $b > 0$, let

$$B_b = \{(\omega_2, \omega_3) : e(\omega_2, \omega_3) \geq b[-\ln(1 - \lambda_n)]^{-1/8}, I(\omega_3) = 1\}.$$

Then by (4.2) and (4.7), if $b = b_\epsilon = \alpha\epsilon^2/2c$,

$$P_{2,3}(B_b) \geq \frac{\alpha}{2}[-\ln(1 - \lambda_n)]^{-9/8}.$$

By an argument as in Corollary 3.6.4, there exists a $\beta = \beta_\epsilon > 0$ such that if

$$\tilde{B} = \{(\omega_2, \omega_3) : e(\omega_2, \omega_3) \geq b[-\ln(1 - \lambda_n)]^{-1/8}, I(\omega_2) = I(\omega_3) = 1\},$$

then

$$P_{2,3}(\tilde{B}) \geq \beta[-\ln(1 - \lambda_n)]^{-9/8}.$$

But, since ω_2 and ω_3 are independent,

$$P_{2,3}(\tilde{B}) \leq [P_3\{\omega_3 : P_1\{\omega_1 \cap \omega_3 = \emptyset\} \geq b[-\ln(1 - \lambda_n)]^{-1/8}, I(\omega_3) = 1\}]^2,$$

and hence

$$P_3\{\omega_3 : P_1\{\omega_1 \cap \omega_3 = \emptyset\} \geq b[-\ln(1 - \lambda_n)]^{-1/8}, I(\omega_3) = 1\} \geq$$

$$\sqrt{\beta}[-\ln(1 - \lambda_n)]^{-9/16}. \tag{4.8}$$

If we let $D(\omega_3) = \{\omega_1 : \omega_1 \cap \omega_3 = \emptyset, I(\omega_1) = 1\}$, then (4.8) implies

$$P\{(\omega_3, \omega_1, \omega_1') : \omega_1, \omega_1' \in D(\omega_3), I(\omega_3) = 1\} \geq$$

$$b^2\sqrt{\beta}[-\ln(1-\lambda_n)]^{-13/16}. \qquad (4.9)$$

Here we are using the fact that $(\Omega_1, P_1) = (\Omega_2, P_2) = (\Omega_3, P_3)$. But it is easy to see that

$$E_{2,3}(eI^+) \geq P\{(\omega_3, \omega_1, \omega_1') : \omega_1, \omega_1' \in D(\omega_3), I(\omega_3) = 1\},$$

so (4.5) and (4.9) give a contradiction. Therefore (4.6) must hold.

To finish the upper bound, (4.4) and (4.6) imply

$$E_{2,3}(I^+) \leq [-(1-\epsilon)4a_4\ln(1-\lambda)]^{-1} + o_\epsilon([-\ln(1-\lambda)]^{-1}).$$

Since this hold for every $\epsilon > 0$,

$$\limsup_{\lambda \to 1-}[-\ln(1-\lambda)]^{-1}E_{2,3}(eI^+) \leq 4a_4.$$

For the lower bound, note that Hölder's inequality and (4.7) give

$$
\begin{aligned}
E_{2,3}(eI^+GY) &\leq [E_{2,3}(G^9)]^{1/9}[E_{2,3}(eI^+Y)]^{8/9}\\
&\leq o_\epsilon([-\ln(1-\lambda)]^{-1})[E_{2,3}(G^9)]^{1/9}.
\end{aligned}
$$

A routine (but somewhat tedious) calculation gives

$$E_{2,3}(G^9) = O([\ln(1-\lambda)]^9).$$

Therefore $E_{2,3}(eI^+GY) = o_\epsilon(1)$ and by (4.1)

$$
\begin{aligned}
1 - o_\epsilon(1) &= E_{2,3}(eI^+G(1-Y))\\
&\leq -(1+\epsilon)4a_4\ln(1-\lambda)E_{2,3}(eI^+).
\end{aligned}
$$

Since this holds for every $\epsilon > 0$,

$$\liminf_{\lambda \to 1-} -[\ln(1-\lambda)]^{-1}E_{2,3}(eI^+) \geq 4a_4. \quad \square$$

Using Proposition 3.6.3 and Lemma 3.2.4 we get two immediate corollaries.

Corollary 4.2.3 *As $\lambda \to 1-$,*

$$F(\lambda) = E_{2,3}(e) \asymp -[\ln(1-\lambda)]^{-1}.$$

Corollary 4.2.4 (Theorem 3.5.1)

$$F(n) \asymp (\ln n)^{-1}.$$

In the next section we will derive asymptotic expressions for the probability of long-range intersections by using results about two-sided intersections. We will need to consider random walks with a fixed number of steps. Let A_n and B_n be the events

$$A_n = \{S^1(0,n] \cap (S^2[0,n] \cup S^3[0,n]) = \emptyset, 0 \notin S^3(0,n],$$

$$B_n = \{S^1(0,n] \cap (S^2[0,\infty] \cup S^3[0,n]) = \emptyset, 0 \notin S^3(0,n]\}.$$

Corollary 4.2.5

$$P(A_n) \sim (4a_4 \ln n)^{-1} = \frac{\pi^2}{8}(\ln n)^{-1}, \qquad (4.10)$$

$$P(B_n) \sim (4a_4 \ln n)^{-1} = \frac{\pi^2}{8}(\ln n)^{-1}. \qquad (4.11)$$

Proof. The result for $P(A_n)$ follows immediately from Theorem 4.2.2 and Lemma 3.2.4. Clearly $P(B_n) \leq P(A_n)$. To prove the other inequality let

$$V_n = \{S^1(0,n] \cap (S^2[0,n\ln n) \cup S^3[0,n)) = \emptyset, 0 \notin S^3[0,n)\},$$

$$W_n = \{S^1[0,n] \cap S^2[n\ln n,\infty) \neq \emptyset\}.$$

Then $B_n = V_n \setminus W_n$. By (3.9),

$$
\begin{aligned}
P(W_n) &\leq c(\ln n)^{-1} \sum_{i=0}^{2n} \sum_{j=n\ln n}^{\infty} P\{S_i^1 = S_j^2\} \\
&\leq c(\ln n)^{-1} \sum_{i=0}^{2n} \sum_{j=n\ln n}^{\infty} (i+j)^{-2} \\
&\leq c(\ln n)^{-2}.
\end{aligned}
$$

Therefore,

$$P(B_n) \geq P(V_n) - P(W_n) \geq P(A_{n\ln n}) - P(W_n) \sim \frac{\pi^2}{8}(\ln n)^{-1}. \quad \square$$

4.3 Long-range Intersections

Let S^1, S^2 be independent simple random walks in Z^4 starting at x and 0 respectively and let R_n be the number of intersections if S^1 takes n steps and S^2 takes an infinite number of steps, i.e.,

$$R_n = \sum_{j=0}^{n} \sum_{k=0}^{\infty} I\{S_j^1 = S_k^2\}.$$

Let $J_n(x) = J(n, x)$ be the expected number of intersections, i.e.,

$$J_n(x) = E^{x,0}(R_n) = \sum_{j=0}^{n} \sum_{k=0}^{\infty} p_{j+k}(x)$$

$$= \sum_{j=0}^{n-1} j p_j(x) + \sum_{j=n}^{\infty} n p_j(x).$$

It is routine to estimate $J_n(x)$ using the local central limit theorem. We state here without proof the results we will need.

Proposition 4.3.1 *If $\alpha > 0$, then as $n \to \infty$,*
  (i)

$$J_n(0) \sim \frac{4}{\pi^2} \ln n;$$

(ii) if $|x_n|^2 \sim n(\ln n)^{-\alpha}$,

$$J_n(x_n) \sim \frac{4\alpha}{\pi^2} \ln \ln n;$$

(iii) if $|x_n|^2 \sim \alpha n$,

$$J_n(x_n) \sim \frac{2}{\alpha \pi^2}(1 - e^{-2\alpha}) + \frac{4}{\pi^2} \int_{2\alpha}^{\infty} \frac{e^{-u}}{u} du;$$

(iv) if $|x_n|^2 \sim n(\ln n)^{\alpha}$,

$$J_n(x_n) \sim \frac{2}{\pi^2}(\ln n)^{-\alpha}.$$

The goal of this section is to give asymptotic expressions for the probability that $S^1[0, n] \cap S^2[0, \infty) \neq \emptyset$. Let

$$\phi_n(x) = P^{x,0}\{S^1[0, n] \cap S^2[0, \infty) \neq \emptyset\}.$$

We define stopping times

$$\begin{aligned}
\tau &= \inf\{j \geq 0 : S_j^1 \in S^2[0, \infty)\} \\
\sigma &= \inf\{k \geq 0 : S_k^2 = S_\tau^1\}.
\end{aligned}$$

Then if $\Lambda(j, k, x) = P^{x,0}\{\tau = j, \sigma = k\}$,

$$\phi_n(x) = P^{x,0}\{\tau \leq n\} = \sum_{j=0}^{n} \sum_{k=0}^{\infty} \Lambda(j, k, x).$$

If we translate so that the origin is at $S_\tau^1 = S_\sigma^2$ and reverse direction on the two paths of finite length, one can easily check that

$$\Lambda(j, k, x) = P\{S^1(0, j] \cap (S^2[0, \infty) \cup S^3[0, k]) = \emptyset,$$

$$0 \notin S^3(0, k], S_j^1 - S_k^3 = x\}.$$

Corollary 4.2.5 gives an estimate of the probability of

$$\{S^1(0, j] \cap (S^2[0, \infty) \cup S^3[0, k]) = \emptyset, 0 \notin S^3(0, k]\}.$$

What we would like to say is that the event $\{S_j^1 - S_k^3 = x\}$ is almost independent of the above event. This will not be true if x is near the origin. However, we will be able to prove this for x sufficiently far from the origin. The next proposition gives an upper bound.

Proposition 4.3.2 *For every $\alpha > 0$, if*

$$n(\ln n)^{-\alpha} \leq |x|^2, j, k \leq n(\ln n)^{\alpha},$$

then

$$\Lambda(j, k, x) \leq \frac{\pi^2}{8}(\ln n)^{-1} p_{j+k}(x)(1 + o_\alpha(1)).$$

Proof. We will assume that $x \leftrightarrow j + k$, i.e., $p_{j+k}(x) > 0$. Let $\beta = \alpha + 2$ and consider the event $A = A_{n(\ln n)^{-\beta}}$ as in Corollary 4.2.5. Then

$$P(A) = \frac{\pi^2}{8}(\ln n)^{-1}(1 + o_\alpha(1)).$$

Let $D = D_{n,\beta}$ be the event

$$\{|S^1([n(\ln n)^{-\beta}])|^2, |S^3([n(\ln n)^{-\beta}])|^2 \leq n(\ln n)^{-2\alpha}\}.$$

Then by Lemma 1.5.1,

$$P(D^c) = O(\exp\{-(\ln n)^2\}) = o(n^{-3}).$$

By the strong Markov property and Theorem 1.2.1,

$$P\{S_j^1 - S_k^3 = x \mid A \cap D\} = p_{j+k}(x)(1 + o_\alpha(1)).$$

Therefore,

$$
\begin{aligned}
\Lambda(j,k,x) &\le P(A \cap \{S_j^1 - S_k^3 = x\}) \\
&\le P(A \cap D \cap \{S_j^1 - S_k^3 = x\}) + P(D^c) \\
&= \frac{\pi^2}{8}(\ln n)^{-1}p_{j+k}(x)(1 + o_\alpha(1)) + o(n^{-3}) \\
&= \frac{\pi^2}{8}(\ln n)^{-1}p_{j+k}(x)(1 + o_\alpha(1)). \quad \square
\end{aligned}
$$

From the above proposition we can get the upper bound for the probability of intersection.

Theorem 4.3.3 *For every $\alpha > 0$, if*

$$n(\ln n)^{-\alpha} \le |x|^2 \le n(\ln n)^\alpha,$$

then

$$
\begin{aligned}
P^{x,0}\{S^1[0,n] \cap S^2[0,\infty) \ne \emptyset\} &\le \frac{J_n(x)}{2J_n(0)}(1 + o_\alpha(1)) \\
&= \frac{\pi^2}{8}(\ln n)^{-1}J_n(x)(1 + o_\alpha(1)).
\end{aligned}
$$

Proof. By Proposition 4.3.1,

$$J_n(x) \ge c(\ln n)^{-\alpha}.$$

By (3.9) and Proposition 4.3.1,

$$
\begin{aligned}
P^{0,x}\{\tau < n(\ln n)^{-3\alpha}\} &= P^{0,x}\{S^1[0, n(\ln n)^{-3\alpha}] \cap S^2[0,\infty) \ne \emptyset\} \\
&\le c_\alpha(\ln n)^{-1}J(n(\ln n)^{-3\alpha}, x) \\
&\le c_\alpha(\ln n)^{-2\alpha-1}.
\end{aligned}
$$

Similarly,

$$
\begin{aligned}
P^{0,x}\{\sigma < n(\ln n)^{-3\alpha}\} &\le c_\alpha(\ln n)^{-2\alpha-1} \\
P^{0,x}\{\sigma > n(\ln n)^{3\alpha}\} &\le c_\alpha(\ln n)^{-2\alpha-1}.
\end{aligned}
$$

Therefore by Proposition 4.3.2

$$P^{x,0}\{S^1[0,n] \cap S^2[0,\infty) \neq \emptyset\}$$

$$\leq \quad c_\alpha(\ln n)^{-2\alpha-1} + \sum_{j=n(\ln n)^{-3\alpha}}^{n} \sum_{k=n(\ln n)^{-3\alpha}}^{n(\ln n)^{3\alpha}} \Lambda(j,k,x)$$

$$\leq \quad c_\alpha(\ln n)^{-2\alpha-1} + (1+o_\alpha(1))\frac{\pi^2}{8}(\ln n)^{-1} \sum_{j=0}^{n} \sum_{k=0}^{\infty} p_{j+k}(x)$$

$$= \quad c_\alpha(\ln n)^{-2\alpha-1} + (1+o_\alpha(1))\frac{\pi^2}{8}(\ln n)^{-1} J_n(x)$$

$$\leq \quad (1+o_\alpha(1))\frac{\pi^2}{8}(\ln n)^{-1} J_n(x). \quad \square$$

We will give a similar lower bound on the probability of intersection if $|x|^2 \asymp n$.

Lemma 4.3.4 *For every $\alpha > 0$, there exists a $c = c_\alpha$ such that if*

$$\alpha n \leq |x|^2, |y|^2 \leq \alpha^{-1}n,$$

then

$$|\phi_n(y) - \phi_n(x)| \leq c|y-x|n^{-1/2}(\ln n)^{-1}.$$

Proof. We will prove the lemma for $y - x = e, |e| = 1$. The general case can be obtained by the triangle inequality. We first estimate $|\Delta\phi_n(x)|$. Note that

$$-\Delta\phi_n(x) \quad = \quad \phi_n(x) - P^{x,0}\{S^1[0,n] \cap S^2(0,\infty) \neq \emptyset\}$$
$$= \quad P^{x,0}\{S^1[0,n] \cap S^2(0,\infty) = \emptyset, 0 \in S^1[0,n]\}.$$

Let $\tau = \inf\{j : S_j^1 = 0\}$. Then by the local central limit theorem,

$$P\{\tau \leq n(\ln n)^{-1} \text{ or } n - n(\ln n)^{-1} \leq \tau \leq n\}$$

$$\leq \quad \sum_{j=0}^{n(\ln n)^{-1}} p_n(x) + \sum_{j=n-n(\ln n)^{-1}}^{n} p_n(x)$$

$$\leq \quad c_\alpha n^{-1}(\ln n)^{-1}.$$

However, if $n(\ln n)^{-1} \leq j \leq n - n(\ln n)^{-1}$ and B_j is the event

$$\{S^1(j) = 0, S^2(0,\infty) \cap (S^1[j,n] \cup S^1[0,j)) = \emptyset, 0 \notin S^1[0,j)\},$$

then an argument as in Lemma 4.3.2 gives

$$P^{x,0}(B_j) \le c_\alpha p_j(x)(\ln n)^{-1}.$$

Therefore, if $\alpha n \le |x|^2 \le \alpha^{-1} n$,

$$
\begin{aligned}
|\Delta\phi_n(x)| \;\le\;& \sum_{j=0}^{\infty} P^{x,0}(B_j) \\
\le\;& c_\alpha n^{-1}(\ln n)^{-1} + c_\alpha(\ln n)^{-1} \sum_{j=n(\ln n)^{-1}}^{n-n(\ln n)^{-1}} p_j(x) \\
\le\;& c_\alpha n^{-1}(\ln n)^{-1}.
\end{aligned}
$$

Let

$$C = C_n = \{z : |z|^2 < \frac{\alpha}{2} n\},$$

$$C_x = C_{n,x} = \{z : |z - x|^2 < \frac{\alpha}{2} n\}.$$

By Theorem 1.4.6 and Exercise 1.5.11, if

$$\sigma = \inf\{j : S_j^1 \in \partial C_x\},$$

then

$$
\begin{aligned}
\phi_n(x) \;=\;& E^x(\phi(S_\sigma)) - \sum_{z \in C_x} \Delta\phi_n(z) G_{C_x}(x, z) \\
=\;& E^x(\phi(S_\sigma)) - \sum_{z \in C} \Delta\phi_n(z + x) G_C(0, z).
\end{aligned}
$$

Since $\psi(z) = E^z(\phi(S_\sigma))$ is harmonic in C_x, Theorem 1.7.1 gives

$$
\begin{aligned}
|E^{x+e}(\phi(S_\sigma)) - E^x(\phi(S_\sigma))| \;\le\;& c_\alpha n^{-1/2} \sup_{z \in \overline{C}_x} |\phi_n(z)| \\
\le\;& c_\alpha n^{-1/2}(\ln n)^{-1}.
\end{aligned}
$$

By "summing by parts" one can verify that

$$\sum_{z \in C}(\Delta\phi_n(x + z + e) - \Delta\phi_n(x + z))G_C(0, z) =$$

$$\sum_{z \in C} \Delta\phi_n(x + z)(G_C(0, z - e) - G_C(0, z))$$

$$+ \sum_{z \in \partial C, z+e \in C} \Delta \phi_n(x+y) G_C(0, z+e).$$

(The other boundary term disappears since $G_C(0, z) = 0$ for $z \notin C$.) If $z \in \partial C$, then by Proposition 1.5.9, $|G_C(0, z+e)| \le c_\alpha n^{-3/2}$. Also, $|\partial C| = O_\alpha(n^{3/2})$, and hence

$$| \sum_{z \in \partial C, z+e \in C} \Delta \phi_n(x+y) G_C(0, z+e)|$$

$$\le \quad c_\alpha n^{3/2} (\sup_{z \in \partial C} |\Delta \phi_n(x)|)(c_\alpha n^{-3/2})$$

$$\le \quad c_\alpha n^{-1} (\ln n)^{-1}.$$

It also follows from Proposition 1.5.9 that for $z \in C$,

$$|G_C(0, z-e) - G_C(0, z)| \le c_\alpha |z|^{-3}.$$

Therefore,

$$\sum_{z \in C} |\Delta \phi_n(x+z)||G_C(0, z-e) - G_C(0, z)|$$

$$\le \quad c_\alpha n^{-1} (\ln n)^{-1/2} \sum_{z \in C} (|z|^{-3} \wedge 1)$$

$$\le \quad c_\alpha n^{-1/2} (\ln n)^{-1/2},$$

which completes the lemma. $\square$

Theorem 4.3.5 *If* $\alpha n \le |x|^2 \le \alpha^{-1} n$,

$$P^{x,0}\{S^1[0, n] \cap S^2[0, \infty) \ne \emptyset\} \quad = \quad \frac{J_n(x)}{2 J_n(0)} (1 + o_\alpha(1))$$

$$= \quad \frac{\pi^2}{8} (\ln n)^{-1} J_n(x)(1 + o_\alpha(1)).$$

Proof. By Theorem 4.3.3,

$$P^{x,0}\{S^1[0, n] \cap S^2[0, \infty) \ne \emptyset\} \le \frac{\pi^2}{8} (\ln n)^{-1} J_n(x)(1 + o_\alpha(1)).$$

If $n(\ln n)^{-1} \le j, k \le n(\ln n)$, then by (4.11),

$$\sum_{y \in Z^4} \Lambda(j, k, y) = \frac{\pi^2}{8} (\ln n)^{-1} (1 + o(1)).$$

If $D = D_n = \{y : n(\ln n)^{-3} \leq |y|^2 \leq n(\ln n)^3\}$, then for such j, k,

$$\sum_{y \notin D} \Lambda(j, k, y) \leq \sum_{y \notin D} p_{j+k}(y) = o((\ln n)^{-1}).$$

Therefore,

$$\sum_{y \in D} \Lambda(j, k, y) = \frac{\pi^2}{8}(\ln n)^{-1}(1 + o(1)).$$

Let $\epsilon > 0$ and let

$$C = C_{\epsilon, n, x} = \{y : |x - y|^2 \leq \epsilon^2 n\}.$$

By Proposition 4.3.3,

$$\sum_{y \in D \setminus C} \Lambda(j, k, y) \leq \sum_{y \in D \setminus C} \frac{\pi^2}{8}(\ln n)^{-1} p_{j+k}(x)(1 + o(1)).$$

Hence,

$$\sum_{y \in C} \Lambda(j, k, y) \geq$$

$$\frac{\pi^2}{8}(\ln n)^{-1}(1 + o(1))(\sum_{y \in C} p_{j+k}(y) - o_\alpha(1) \sum_{y \in D \setminus C} p_{j+k}(y)).$$

If we sum over $n(\ln n)^{-1} \leq j, k \leq n(\ln n)$ and use Proposition 4.3.3 we get

$$\sum_{y \in C} \phi_n(y) = (1 + o_\epsilon(1)) \sum_{y \in C} J_n(y) \frac{\pi^2}{8}(\ln n)^{-1}.$$

By Lemma 4.3.4,

$$|\sup_{y \in C} \phi_n(y) - \inf_{y \in C} \phi_n(y)| \leq c_\alpha \epsilon (\ln n)^{-1}.$$

Therefore,

$$\begin{aligned}
\phi_n(x) + c_\alpha \epsilon (\ln n)^{-1} &\geq \sup_{y \in C} \phi_n(y) \\
&\geq \frac{1}{|C|} \sum_{y \in C} \phi_n(y) \\
&\geq \frac{1}{|C|}(1 + o_\epsilon(1)) \frac{\pi^2}{8}(\ln n)^{-1} \sum_{y \in C} J_n(y) \\
&= (1 + o_\epsilon(1)) \frac{\pi^2}{8}(\ln n)^{-1} J_n(x)(1 + \delta(\epsilon)),
\end{aligned}$$

where $\delta(\epsilon) \to 0$ as $\epsilon \to 0$. The last step uses Proposition 4.3.1(iii). Since this holds for every $\epsilon > 0$, the theorem follows. $\quad\Box$

Similar results hold for long-range intersections of two random walks starting at the origin. Let $0 < a < b < \infty$ and consider the intersections of $S^1[an, bn]$ with $S^2[0, \infty)$, where S^1 and S^2 both start at the origin. The expected number of intersections is given by

$$\sum_{j=an}^{bn}\sum_{k=0}^{\infty} P\{S_j^1 = S_k^2\} \quad = \quad \sum_{j=an}^{bn}\sum_{k=j}^{\infty} p_{j+k}(0)$$

$$\sim \quad \sum_{j=an}^{bn} \frac{4}{\pi^2 j} \sim \frac{4}{\pi^2}\ln(b/a).$$

Similarly if $\alpha > 0$, the expected number of intersections of $S^1[n(\ln n)^{-\alpha}, n]$ and $S^2[0, \infty)$ is given by

$$\sum_{j=n(\ln n)^{-\alpha}}^{n}\sum_{k=0}^{\infty} p_{j+k}(0) \sim \sum_{j=n(\ln n)^{-\alpha}}^{n} \frac{4}{\pi^2 j} \sim \frac{4\alpha}{\pi^2}\ln\ln n.$$

The following theorem about the probability of intersection can be proved in the same way as Theorems 4.3.5 and 4.3.3.

Theorem 4.3.6 *(i) If $0 < a < b < \infty$, then*

$$P\{S^1[an, bn] \cap S^2(0, \infty) \neq \infty\} \quad \sim \quad \frac{1}{2}[\frac{4}{\pi^2}\ln(b/a)][J_n(0)]^{-1}$$

$$= \quad \frac{1}{2}\ln(b/a)(\ln n)^{-1}.$$

(ii) If $\alpha > 0$, then

$$P\{S^1[n(\ln n)^{-\alpha}, n] \cap S^2[0, \infty) \neq \emptyset\}$$

$$\leq \quad \frac{1}{2}(\frac{4\alpha}{\pi^2}\ln\ln n)[J_n(0)]^{-1}(1 + o_\alpha(1))$$

$$= \quad \frac{\alpha}{2}\frac{\ln\ln n}{\ln n}(1 + o_\alpha(1)).$$

4.4 One-sided Walks

We return to the problem of intersections of two walks starting at the origin in Z^4, i.e.,

$$f(n) = P\{S^1(0, n] \cap S^2(0, n] = \emptyset\}.$$

Since we have now proved Theorem 3.5.1 for $d = 4$ we can conclude from (3.30) that

$$c_1(\ln n)^{-1} \leq f(n) \leq c_2(\ln n)^{-1/2}.$$

The goal of this section is to show that the right hand inequality is almost sharp. More specifically we prove the following theorem.

Theorem 4.4.1 *If $d = 4$,*

$$f(n) \approx (\ln n)^{-1/2}. \tag{4.12}$$

Let us first motivate why (4.12) should be true. It will be easier to consider

$$\overline{f}(n) = P\{S^1(0, n] \cap S^2[0, \infty) = \emptyset\}.$$

Clearly, $\overline{f}(n) \leq f(n)$. We define events

$$\overline{A}_n = \{S^1(0, n] \cap S^2[0, \infty) = \emptyset\},$$

$$\overline{B}_n = \{S^1(n, 2n] \cap S^2[0, \infty) = \emptyset\}.$$

Then $\overline{A}_{2n} = \overline{A}_n \cap \overline{B}_n$ and

$$\overline{f}(2n) = P(\overline{A}_n)P(\overline{B}_n \mid \overline{A}_n).$$

By Theorem 4.3.6(i),

$$1 - P(\overline{B}_n) \sim \frac{\ln 2}{2 \ln n}.$$

Intuitively one would expect that paths which have not intersected up through time n would be farther apart than paths which have intersected and hence would be less likely to intersect in the next n steps. In other words, one would guess that $\overline{A}_n$ and $\overline{B}_n$ were positively correlated, i.e., $P(\overline{B}_n \mid \overline{A}_n) \geq P(\overline{B}_n)$. If this were the case we would have

$$\overline{f}(2n) \geq \overline{f}(n)(1 - \frac{\ln 2}{2 \ln n}\rho_n), \tag{4.13}$$

where $\rho_n \to 1$. What we will show in the next proposition is that (4.13) implies (4.12). Note that if $\psi(n) = (\ln n)^{-1/2}$,

$$\psi(2n) \sim \psi(n)(1 - \frac{\ln 2}{2 \ln n}).$$

Proposition 4.4.2 *If $\overline{f}$ is a decreasing positive function satisfying (4.13) for some $\rho_n \to 1$, then for every $\epsilon > 0$,*

$$\liminf_{n \to \infty} (\ln n)^{\frac{1}{2}+\epsilon}\overline{f}(n) = \infty.$$

Proof. Let $\epsilon \in (0,1)$ and choose M so that $|1 - \rho_n| < \epsilon$ for $n \geq M$. Let

$$g(k) = \ln \overline{f}(2^k M).$$

Then (4.13) becomes

$$g(k+1) \geq g(k) + \ln(1 - \frac{\rho_n}{2(k + \log_2 M)}),$$

and if $k > 1$,

$$g(k) \geq g(1) + \sum_{j=1}^{k-1} \ln(1 - \frac{1+\epsilon}{2j}).$$

For j sufficiently large, $\ln(1 - \frac{1+\epsilon}{2j}) \geq -(\frac{1}{2} + \frac{2\epsilon}{3})j^{-1}$. Therefore,

$$\limsup_{k \to \infty} (\ln k)^{-1} g(k) \geq -(\frac{1}{2} + \frac{2\epsilon}{3}),$$

or,

$$\limsup_{k \to \infty} \frac{\ln \overline{f}(2^k M)}{\ln \ln 2^k M} \geq -(\frac{1}{2} + \frac{2\epsilon}{3}).$$

This implies

$$\liminf_{k \to \infty} (\ln 2^k M)^{\frac{1}{2}+\epsilon} \overline{f}(2^k M) = \infty.$$

Since $\overline{f}$ is decreasing, one then easily derives the proposition.   □

The problem with using the above argument to conclude (4.12) is that one cannot prove (4.13), i.e., it is very difficult to show that the events $\overline{A}_n$ and $\overline{B}_n$ are positively correlated. One expects, in fact, that the events are asymptotically independent. It turns out that one can show a sufficient form of asymptotic independence if one works with increments which are logarithmic multiples rather than multiples of 2. Fix $\alpha > 0$ and let

$$A_n = A_{n,\alpha} = \{S^1(0, n(\ln n)^{-\alpha}] \cap S^2[0, \infty) = \emptyset\},$$

$$B_n = B_{n,\alpha} = \{S^1(n(\ln n)^{-\alpha}, n] \cap S^2[0, \infty) = \emptyset\}.$$

Then by Theorem 4.3.6(ii),

$$1 - P(B_n) \leq \frac{\alpha}{2} \frac{\ln \ln n}{\ln n} (1 + o_\alpha(1)).$$

Again, $\overline{f}(n) = P(A_n)P(B_n \mid A_n)$. If it were true that

$$1 - P(B_n \mid A_n) \leq (1 - P(B_n))(1 + o_\alpha(1)), \qquad (4.14)$$

then we would have

$$\overline{f}(n) \geq \overline{f}(n(\ln n)^{-\alpha})(1 - \frac{\alpha}{2}\frac{\ln \ln n}{\ln n}\rho_n), \qquad (4.15)$$

for some $\rho_n \to 1$. That (4.15) suffices to prove (4.12) follows from the following proposition which is proved like Proposition 4.4.2.

Proposition 4.4.3 *If $\overline{f}$ is a decreasing positive function satisfying (4.15) for some $\alpha > 0$ and some $\rho_n \to 1$, then for every $\epsilon > 0$,*

$$\liminf_{n\to\infty}(\ln n)^{\frac{1}{2}+\epsilon}\overline{f}(n) = \infty.$$

We have therefore reduced Theorem 4.4.1 to showing that for some $\alpha > 0$,

$$P(B_n^c \mid A_n) \leq P(B_n^c)(1 + o(1)).$$

The next proposition gives a further reduction.

Proposition 4.4.4 *Suppose for some $\alpha > 0$,*

$$P(B_n^c \mid A_{n(\ln n)^{-\alpha}}) \sim P(B_n^c), \qquad (4.16)$$

then

$$P(B_n^c \mid A_n) \leq P(B_n^c)(1 + o(1)). \qquad (4.17)$$

Proof. We first recall from (4.11) that

$$\overline{f}(n) = P(A_n) \geq c(\ln n)^{-1}. \qquad (4.18)$$

For any n, i, let

$$D^i = D_n^i = \{S^1(n(\ln n)^{-\alpha i}, n(\ln n)^{-\alpha(i-1)}] \cap S^2[0, \infty) \neq \emptyset\}.$$

$$A^i = A_n^i = A_{n(\ln n)^{-\alpha i}}.$$

Note that $B_n^c = D_n^1$. Suppose (4.17) does not hold, i.e., for some $\epsilon > 0$,

$$\limsup_{n\to\infty}\frac{\ln n}{\ln \ln n}P(B_n^c \mid A_n) \geq \frac{1}{2}(\alpha + 3\epsilon). \qquad (4.19)$$

Choose a large n with

$$P(D^1 \mid A^1) \geq \frac{1}{2}(\alpha + 2\epsilon)\frac{\ln \ln n}{\ln n}.$$

By assumption, if n is sufficiently large, $i = 1, 2, 3, 4$,

$$P(D^i \mid A^{i+1}) \leq \frac{1}{2}(\alpha + \epsilon)\frac{\ln \ln n}{\ln n}.$$

Note that

$$P(D^i \mid A^{i+1}) \geq P(D^i \cap A^i \mid A^{i+1}) = P(D^i \mid A^i)P(A^i \mid A^{i+1}),$$

or

$$P(A^i \mid A^{i+1}) \leq \frac{P(D^i \mid A^{i+1})}{P(D^i \mid A^i)}.$$

If we consider $i = 1$ we get

$$P(A^1 \mid A^2) \leq \frac{\alpha + \epsilon}{\alpha + 2\epsilon},$$

or

$$P(D^2 \mid A^2) = 1 - P(A^1 \mid A^2) \geq \frac{\epsilon}{\alpha + 2\epsilon}.$$

For $i = 2$ we get

$$P(A^2 \mid A^3) \leq \frac{P(D^2 \mid A^3)}{P(D^2 \mid A^2)} \leq \frac{\ln\ln n}{\ln n} \frac{(\alpha + \epsilon)(\alpha + 2\epsilon)}{2\epsilon},$$

and hence if n is sufficiently large,

$$P(D^3 \mid A^3) = 1 - P(A^2 \mid A^3) \geq \frac{1}{2}.$$

We can iterate again giving

$$P(A^3 \mid A^4) \leq \frac{P(D^3 \mid A^4)}{P(D^3 \mid A^3)} \leq \frac{\ln\ln n}{\ln n}(\alpha + \epsilon).$$

But

$$
\begin{aligned}
P(A^2) &\leq P(A^2 \mid A^4) \\
&= P(A^2 \mid A^3)P(A^3 \mid A^4) \\
&\leq (\frac{\ln\ln n}{\ln n})^2 \frac{(\alpha + \epsilon)^2(\alpha + 2\epsilon)}{2\epsilon}.
\end{aligned}
$$

But this cannot hold for arbitrarily large n by (4.18). Therefore (4.19) cannot hold, and the proposition follows. $\square$

We are now in a position to prove Theorem 4.4.1. From the above if suffices to prove (4.16) for $\alpha = 9$. Let

$$a_n = [n(\ln n)^{-9}][\ln(n(\ln n)^{-9})]^{-9} \sim n(\ln n)^{-18},$$

$$b_n = n(\ln n)^{-16}, c_n = n(\ln n)^{-11}, d_n = n(\ln n)^{-9},$$

and define events

$$
\begin{aligned}
\overline{V}_n &= \{S^1(0, a_n] \cap S^2[0, \infty) = \emptyset\}, \\
\tilde{V}_n &= \{S^1(0, a_n] \cap S^2[0, b_n] = \emptyset\}, \\
V_n &= \tilde{V}_n \cap \{|S^1(b_n)|^2 \le n(\ln n)^{-13}, |S^3(b_n)|^2 \le n(\ln n)^{-13}\}, \\
\overline{W}_n &= \{S^1[d_n, n] \cap S^2[0, \infty) \ne \emptyset\}, \\
W_n &= \{S^1[d_n, n] \cap S^2[c_n, \infty) \ne \emptyset\}.
\end{aligned}
$$

In this notation, (4.16) becomes

$$
P(\overline{W}_n) = P(\overline{W}_n \mid \overline{V}_n)(1 + o(1)). \tag{4.20}
$$

We know by (4.18) that

$$
P(\overline{V}_n) \ge c(\ln n)^{-1}. \tag{4.21}
$$

By Proposition 4.3.1(iv),

$$
P(\tilde{V}_n^c \cap \overline{V}_n) \le P\{S^1(0, a_n] \cap S^2[b_n, \infty) \ne \emptyset\} \le O((\ln n)^{-2}). \tag{4.22}
$$

Similarly,

$$
P(\overline{W}_n^c \cap W_n) \le O((\ln n)^{-2}). \tag{4.23}
$$

By Lemma 1.5.1,

$$
P\{|S^1(b_n)|^2 \ge n(\ln n)^{-13}\} = o((\ln n)^{-2}),
$$

and hence

$$
P(V_n^c \mid \tilde{V}_n) = o((\ln n)^{-2}). \tag{4.24}
$$

Since $P(\overline{W}_n) \ge O((\ln n)^{-1})$ (Theorem 4.3.6(i)), it follows from (4.21)-(4.24) that to prove (4.20) it suffices to prove

$$
P(W_n) = P(W_n \mid V_n)(1 + o(1)). \tag{4.25}
$$

Let

$$
\phi(x, y) = \phi_n(x, y) = P^{x,y}\{S^1[d_n - b_n, \infty) \cap S^2[c_n - b_n, \infty) = \emptyset\}.
$$

Then by the strong Markov property,

$$
P(W_n \mid V_n) = E(\phi(S^1(b_n), S^2(b_n)) \mid V_n).
$$

It suffices, therefore, to show that for $|x|^2, |y|^2 \le n(\ln n)^{-13}$,

$$
\phi(x, y) = \phi(0, 0)(1 + o(1)).
$$

This can be done by an argument as in Lemma 4.3.4.

4.5   Three Walks in Three Dimensions

In this section we will consider the mutual intersection of three random walks in three dimensions. As mentioned in the beginning of this chapter, three is the critical dimension for intersections of three walks. The results are analogous to the results for two walks in four dimensions and the methods of proof are very similar [41], so we will only give some of the main results without proofs. Let S^1, S^2, S^3 be independent simple random walks in Z^3 starting at the origin and let $R_{n,m} = R(n,m)$ be the number of mutual intersections of $S^1[n, m], S^2[0, \infty], S^3[0, \infty)$. i.e.,

$$R_{n,m} = \sum_{i=n}^{m} \sum_{j=0}^{\infty} \sum_{k=0}^{\infty} I\{S_i^1 = S_j^2 = S_k^3\}.$$

If $J_{n,m} = E(R_{n,m})$, then

$$
\begin{aligned}
J_{n,m} &= \sum_{i=n}^{m} \sum_{j=0}^{\infty} \sum_{k=0}^{\infty} P\{S_i^1 = S_j^2 = S_k^3\} \\
&= \sum_{i=n}^{m} E(G(S_i^1)^2),
\end{aligned}
$$

where G is the usual Green's function on Z^3. By Theorem 1.5.4, as $i \to \infty$,

$$E(G(S_i^1)^2) \sim a_3^2 E(|S_i^1|^{-2} \wedge 1).$$

By the central limit theorem, if X is a standard normal random variable in R^3,

$$E(|S_i^1|^2 \wedge 1) \sim 3i^{-1} E(|X|^{-2}) = 3i^{-1}.$$

Therefore,

$$J_{0,n} \sim 3a_3^2 \ln n, \tag{4.26}$$

$$J_{an,bn} \sim 3a_3^2 \ln(b/a). \tag{4.27}$$

Suppose $S^1[an, bn] \cap S^2[0, \infty) \cap S^3[0, \infty) \neq \emptyset$ and let

$$
\begin{aligned}
\tau = \tau_{n,a} &= \inf\{i \geq an : S_i^1 \in S^2[0, \infty) \cap S^3[0, \infty)\}, \\
\sigma = \sigma_{n,a} &= \inf\{j : S_j^2 = S_\tau^1\}, \\
\eta = \eta_{n,a} &= \inf\{k : S_k^3 = S_\tau^1\}.
\end{aligned}
$$

Then

$$P\{S^1[an, bn] \cap S^2[0, \infty) \cap S^3[0, \infty) \neq \emptyset\} =$$

$$\sum_{i=an}^{bn}\sum_{j=0}^{\infty}\sum_{k=0}^{\infty}P\{\tau=i,\sigma=j,\eta=k\}.$$

As in the case of intersections of two walks, by moving the origin we can write the probability of these long-range intersections in terms of probabilities of intersection of walks starting at the origin. Suppose $S^1,\ldots,S^5$ are independent walks starting at the origin. Let $A(i,j,k)$ be the event

$$\{S^1(0,i]\cap(S^2[0,\infty)\cup S^3[0,j])\cap(S^4[0,\infty)\cup S^5[0,k])=\emptyset,$$

$$0\notin S^3(0,j]\cup S^5(0,k]\}.$$

Then

$$P\{\tau=i,\sigma=j,\eta=k\}=P(A(i,j,k)\cap\{S_i^1=S_j^2=S_k^3\}).$$

To estimate $A(i,j,k)$ we first derive a result for killed random walks. Suppose S^1,S^3,S^5 are killed at rate $1-\lambda$ with killing times T^1,T^3,T^5 (it is not necessary to kill S^2 and S^4) and define the following random variables on $\Omega_2\times\Omega_3\times\Omega_4\times\Omega_5$:

$$e=e_\lambda=P_1\{S^1(0,T^1]\cap(S^2[0,\infty)\cup S^3[0,T^3])\cap$$

$$(S^4[0,\infty)\cup S^5[0,T^5))=\emptyset\},$$

$$I=I_\lambda=\text{ indicator function of }\{0\notin S^3(0,T^3]\cup S^5(0,T^5]\},$$

$$G=G^\lambda=\sum_{j_2=0}^{\infty}G_\lambda(S_{j_2}^2)+\sum_{j_3=1}^{T^3}G_\lambda(S_{j_3}^3)+$$

$$\sum_{j_4=0}^{\infty}G_\lambda(S_{j_4}^4)+\sum_{j_5=1}^{T_5}G_\lambda(S_{j_5}^5).$$

Then the following is proved like Theorem 3.6.1.

Theorem 4.5.1

$$E_{2,3,4,5}(eIG)=1.$$

As $\lambda\to1-$, one can show (see (4.26)) that

$$E_{2,3,4,5}(G)\sim-12a_3^2\ln(1-\lambda).$$

The extra factor of 4 comes from the fact that there are 4 walks, $S^2,\ldots,S^5$. As in Lemma 4.2.1(b) one can also show that

$$\text{Var}_{2,3,4,5}(G)=O(-\ln(1-\lambda)),$$

and hence we get this analogue of Theorem 4.2.2.

Theorem 4.5.2 *As $\lambda \to 1-$,*

$$E_{2,3,4,5}(eI) \sim [-12a_3^2 \ln(1-\lambda)]^{-1}.$$

We can then use a standard Tauberian argument to see that if $\alpha > 0$ and $n(\ln n)^{-\alpha} \leq i, j, k \leq n(\ln n)^{\alpha}$,

$$P(A(i,j,k)) \sim [12a_3^2 \ln n]^{-1}.$$

Again we can show for such i, j, k that

$$P(A(i,j,k) \cap \{S_i^1 = S_j^2 = S_k^3\}) \sim [12a_3^2 \ln n]^{-1} P\{S_i^1 = S_j^2 = S_k^3\}$$

and deduce the following.

Theorem 4.5.3 *If $0 < a < b < \infty$,*

$$P\{S^1[an, bn] \cap S^2[0, \infty) \cap S^3[0, \infty) \neq \emptyset\} \sim \frac{1}{4} \ln(b/a)(\ln n)^{-1}.$$

If we let

$$H(n) = P\{S^1(0, n] \cap (S^2[0, \infty) \cup S^3[0, \infty)) \cap (S^4[0, \infty) \cup S^5[0, \infty)) = \emptyset\},$$

$$h(n) = P\{S^1(0, n] \cap S^2[0, \infty) \cap S^3[0, \infty) = \emptyset\},$$

then it can be derived from Theorem 4.5.1 that

$$H(n) \asymp (\ln n)^{-1}.$$

Clearly $h(n) \geq H(n)$. Let

$$r(n) = P\{S^1(0, n] \cap (S^2[0, \infty) \cup S^3[0, \infty)) \cap S^4[0, \infty) = \emptyset\}.$$

Then by Hölder's inequalty (see Section 3.7) one can show

$$r(n) \leq h(n) \leq \sqrt{r(n)},$$

$$H(n) \leq r(n) \leq \sqrt{H(n)}.$$

Therefore,

$$c_1(\ln n)^{-1} \leq h(n) \leq c_2(\ln n)^{-1/4}.$$

It can be shown in fact that the upper bound is almost sharp. Note that by Theorem 4.5.3,

$$P\{S^1(n, 2n] \cap S^2[0, \infty) \cap S^3[0, \infty) \neq \emptyset\} \sim \frac{1}{4}(\ln 2)(\ln n)^{-1}.$$

Suppose that we could show

$$h(2n) \geq h(n)\left(1 - \frac{1}{4}\ln 2(\ln n)^{-1}\rho_n\right), \tag{4.28}$$

where $\rho_n \to 1$. Then by an argument similar to Proposition 4.4.2 we could deduce for any $\epsilon > 0$,

$$\liminf_{n\to\infty}(\ln n)^{\frac{1}{4}+\epsilon}h(n) = \infty.$$

It turns out again that (4.28) is difficult to prove directly. However, one can work with logarithmic multiples as was done in the last section and deduce the following.

Theorem 4.5.4 *If $d = 3$,*

$$P\{S^1(0,n] \cap S^2[0,\infty) \cap S^3[0,\infty) = \emptyset\} \approx (\ln n)^{-1/4}.$$

Chapter 5

Two and Three Dimensions

5.1 Intersection Exponent

In this chapter we study

$$f(n) = P\{S^1(0,n] \cap S^2(0,n] = \emptyset\},$$

where S^1, S^2 are independent simple random walks in Z^2 or Z^3. By (3.29),

$$c_1 n^{(d-4)/2} \leq f(n) \leq c_2 n^{(d-4)/4}, \tag{5.1}$$

so we would expect that

$$f(n) \approx n^{-\zeta},$$

for some $\zeta = \zeta_d$. We show that this is the case and that the exponent is the same as an exponent for intersections of Brownian motions. Let B^1, B^2 be independent Brownian motions in R^d starting at distinct points x, y. It was first proved in [19] that if $d < 4$,

$$P^{x,y}\{B^1[0,\infty) \cap B^2[0,\infty) \neq \emptyset\} = 1.$$

Let

$$b(x,y,r) = P^{x,y}\{B^1[0,T_r^1] \cap B^2[0,T_r^2] = \emptyset\},$$

where

$$T_r^i = \inf\{t : |B^i(t)| = r\}.$$

We prove that as $r \to \infty$,

$$b(x,y,r) \approx r^{-\xi}, \tag{5.2}$$

where $\xi = 2\zeta$. Proving the existence of a ξ satisfying (5.2) is relatively straightforward using subadditivity, but it takes more work to show that $\xi = 2\zeta$, and hence that ζ exists.

The next problem, of course, is to compute ζ or ξ. Unfortunately, this is still an open problem and only partial results can be given. Duplantier [16] gives a nonrigorous renormalization group expansion for ζ in $d = 4 - \epsilon$ which suggests that both bounds in (5.1) are not sharp, i.e.,

$$\frac{d-4}{4} < \zeta < \frac{d-4}{2}.$$

A nonrigorous conformal invariance argument by Duplantier and Kwon [17] gives a conjecture that $\zeta_2 = 5/8$. In analogy with a number of exponents found in mathematical physics, one would expect that ζ_2 would be a rational number (conformal invariance argues for this) while for $d = 3$ the exponent could well be an irrational number that cannot be calculated exactly. Monte Carlo simulations [13, 17, 49] are consistent with $\zeta_2 = 5/8$ and give a value for ζ_3 between .28 and .29.

In this section we prove

$$\frac{1}{2} + \frac{1}{4\pi^2} \leq \zeta_2 < \frac{3}{4}.$$

The lower bound can be improved slightly using a similar argument to give the best known bounds

$$\frac{1}{2} + \frac{1}{8\pi} \leq \zeta_2 < \frac{3}{4}.$$

The lower bound is achieved by comparison to an exponent $\bar{\gamma}$ defined by saying that the probability that a Brownian motion starting at $e_1 = (1,0)$ does not form a closed loop about 0 before hitting the sphere of radius r decays like $r^{-\bar{\gamma}}$. We prove that

$$\zeta_2 \geq \frac{1}{2} + \frac{\bar{\gamma}}{4}, \tag{5.3}$$

and then give bounds on $\bar{\gamma}$. It has been conjectured that $\bar{\gamma} = 1/4$. The upper bound makes use of a standard result on harmonic measures on the unit disk, the Beurling projection theorem [1]. For $d = 3$ we can only prove that the upper bound is not sharp,

$$\frac{1}{4} \leq \zeta_3 < \frac{1}{2}.$$

5.2 Intersections of Brownian Motions

Let B_t^1, B_t^2 be independent Brownian motions in R^d $(d = 2, 3)$. Let D_r be the ball of radius r

$$D_r = \{x \in R^d : |x| < r\},$$

$$\partial D_r = \{x \in R^d : |x| = r\},$$

and let T_r^i be the hitting time of ∂D_r,

$$T_r^i = \inf\{t : |B_t^i| = r\}.$$

We write D for D_1 and T^i for T_1^i. If $|x|, |y| < r, x \neq y$, we define

$$b(x, y, r) = P^{x,y}\{B^1[0, T_r^1] \cap B^2[0, T_r^2] = \emptyset\}.$$

By [19], $0 < b(x, y, r) < 1$ and

$$\lim_{r \to \infty} b(x, y, r) = 0.$$

By scaling properties of Brownian motion, if $c > 0$,

$$b(cx, cy, cr) = b(x, y, r).$$

We let e_1 be the unit vector in R^d whose first component is 1. If $z \in R^d$, we let

$$D_r(z) = \{x + z : x \in D_r\}.$$

If $|x|, |y| < r$,

$$D_{r-\frac{1}{2}|x+y|}\left(\frac{x+y}{2}\right) \subset D_r \subset D_{r+\frac{1}{2}|x+y|}\left(\frac{x+y}{2}\right).$$

Therefore, by translation and rotational invariance of Brownian motion and scaling,

$$b\left(e_1, -e_1, \frac{2r - |x+y|}{|x-y|}\right) \geq b(x, y, r) \geq b\left(e_1, -e_1, \frac{2r + |x+y|}{|x-y|}\right). \tag{5.4}$$

We now investigate the behavior of $b(x, y, r)$ as $r \to \infty$. We will use properties of subadditive functions. A function $\phi : \{0, 1, 2, \ldots\} \to R$ is called subadditive if

$$\phi(j + k) \leq \phi(j) + \phi(k).$$

Lemma 5.2.1 *If ϕ is a subadditive function from $\{0, 1, 2, \ldots\}$ to R, then*

$$\lim_{n \to \infty} \frac{\phi(n)}{n} = \inf_{n > 0} \frac{\phi(n)}{n}.$$

Proof. Clearly

$$\liminf_{n\to\infty} \frac{\phi(n)}{n} \geq \inf_{n>0} \frac{\phi(n)}{n}.$$

To prove the other direction, let m be a positive integer, and $a = a(m) = \sup\{\phi(k) : 0 \leq k \leq m-1\}$. If n is any integer, write $n = jm + k$ where $0 \leq k \leq m-1$. Then

$$\frac{\phi(n)}{n} = \frac{\phi(jm+k)}{n} \leq \frac{j\phi(m) + a}{n} \leq \frac{\phi(m)}{m} + \frac{a}{n}.$$

Therefore,

$$\limsup_{n\to\infty} \frac{\phi(n)}{n} \leq \frac{\phi(m)}{m}. \quad \square$$

Theorem 5.2.2 *There exists a $\xi = \xi_d \leq 2$ such that for $x \neq y$,*

$$\lim_{r\to\infty} -\frac{\ln b(x,y,r)}{\ln r} = \xi.$$

Proof. Let

$$b(r) = \sup_{|x|,|y|=1} b(x,y,r).$$

For $|x|, |y| = 1$, $1 < s < r$, by the strong Markov property and scaling,

$$\begin{aligned}
b(x,y,r) &\leq P^{x,y}\{B^1[0,T_s^1] \cap B^2[0,T_s^2] = \emptyset, \\
&\qquad\qquad B^1[T_s^1, T_r^1] \cap B^2[T_s^2, T_r^2] = \emptyset\} \\
&\leq b(x,y,s)b(\tfrac{r}{s}).
\end{aligned}$$

Therefore $b(r_1 r_2) \leq b(r_1)b(r_2)$ and if we let

$$\phi(k) = \ln b(2^k),$$

then ϕ is a subadditive function. By Lemma 5.2.1, there exists an a (perhaps equal to $-\infty$) such that

$$\lim_{k\to\infty} \frac{\phi(k)}{k} = a,$$

i.e.,

$$\lim_{k\to\infty} -\frac{\ln b(2^k)}{\ln 2^k} = -\frac{a}{\ln 2} \doteq \xi.$$

Since $b(r)$ is decreasing in r, it is easy to see that this implies

$$b(r) \approx r^{-\xi}.$$

By (5.4),

$$b(e_1, -e_1, r) \leq b(r) \leq b(e_1, -e_1, r-1).$$

Therefore,

$$b(e_1, -e_1, r) \approx r^{-\xi}.$$

If we use (5.4) again, we get the result for general x, y.

To see that $\xi \leq 2$, let σ_r^i be the first time that the first component of $B_t^i, [B_t^i]_1$, is equal to $0, r$, or $-r$. By the standard "gambler's ruin" estimate for one-dimensional Brownian motion,

$$P^{\pm e_1}\{[B^i(\sigma_r^i)]_1 = \pm r\} = \frac{1}{r}.$$

But,

$$
\begin{aligned}
b(e_1, -e_1, r) &\geq P^{e_1, -e_1}\{[B^1(\sigma_r^1)]_1 = r, [B^2(\sigma_r^2)]_1 = -r\} \\
&\geq r^{-2},
\end{aligned}
$$

and hence $\xi \leq 2$. □

We will now relate random walk intersections and Brownian motion intersections. If S_j is a random walk in Z^d and $\tau_n = \inf\{j : |S_j| \geq n\}$, then by the invariance principle

$$W_n(t) \doteq n^{-1}S([tn^2]), \quad 0 \leq t \leq n^{-2}\tau_n$$

converges in distribution to a Brownian motion B_t with covariance $d^{-1}I$ (i.e., $B_t = d^{-\frac{1}{2}}\tilde{B}_t$ where $\tilde{B}_t$ is a standard Brownian motion) stopped upon hitting the ball of radius 1. Note that

$$
\begin{aligned}
P^{x,y}&\{B^1[0, T_r^1] \cap B^2[0, T_r^2] = \emptyset\} \\
&= P^{x/\sqrt{d}, y/\sqrt{d}}\{\tilde{B}^1[0, T_{r/\sqrt{d}}^1] \cap \tilde{B}^2[0, T_{r/\sqrt{d}}^2] = \emptyset\} \\
&= b(\frac{x}{\sqrt{d}}, \frac{y}{\sqrt{d}}, \frac{r}{\sqrt{d}}) = b(x, y, r),
\end{aligned}
$$

i.e., the probability of intersection before hitting the sphere of radius r is the same for $\tilde{B}_t$ and B_t. It will be convenient for the next two sections to let B_t be a Brownian motion with covariance $d^{-1}I$. The remainder of this chapter will be devoted to proving the following theorem. Recall that C_n is the "ball" of radius n contained in Z^d, and let τ_n^i be the first hitting time of ∂C_n by S^i,

$$\tau_n^i = \inf\{j : |S_j^i| \geq n\}.$$

Theorem 5.2.3 *If* $d = 2, 3$, $x_n, y_n \in C_n, x, y \in D$ *with* $n^{-1}x_n \to x$, $n^{-1}y_n \to y$, *then*

$$\lim_{n \to \infty} P^{x_n, y_n}\{S^1[0, \tau_n^1] \cap S^2[0, \tau_n^2] = \emptyset\} = b(x, y, 1).$$

The theorem first appears to be a simple consequence of the invariance principle, and one direction can be proved easily this way. Let $\delta > 0$ and

$$A_\delta = \{\text{dist}(B^1[0, T^1], B^2[0, T^2]) < \delta\},$$

$$U_{\delta, n} = \{\text{dist}(S^1[0, \tau_n^1], S^2[0, \tau_n^2]) < n\delta\}.$$

Then it follows from the invariance principle that for every $\delta > 0$,

$$\limsup_{n \to \infty} P^{x_n, y_n}(U_{\delta, n}) \le P^{x, y}(A_{2\delta}),$$

$$\liminf_{n \to \infty} P^{x_n, y_n}(U_{2\delta, n}) \ge P^{x, y}(A_\delta).$$

Note that as $\delta \to 0$, $P^{x, y}(A_\delta) \to 1 - b(x, y, 1)$. Also,

$$\{S^1[0, \tau_n^1] \cap S^2[0, \tau_n^2] \ne \emptyset\} \subset U_{\delta, n}.$$

Therefore,

$$\liminf_{n \to \infty} P^{x_n, y_n}\{S^1[0, \tau_n^1] \cap S^2[0, \tau_n^2] = \emptyset\} \ge b(x, y, 1). \qquad (5.5)$$

To prove the other direction is trickier. Essentially we have to show that if two random walks get "close" to each other, then with high probability they will actually intersect. We will do the proof for $d = 3$; the $d = 2$ case can be done similarly (one can also give some easier proofs in the $d = 2$ case). We start by stating without a proof a proposition which can be proved in a similar fashion to Theorem 3.3.2.

Proposition 5.2.4 *If $d < 4$, for every $r > 1$, there exists a $c = c(r) > 0$ such that if $x, y \in C_n$,*

$$P^{x, y}\{S^1[0, \tau_{rn}^1] \cap S^2[0, \tau_{rn}^2] \ne \emptyset\} \ge c.$$

We will need a slightly stronger version of this proposition.

Proposition 5.2.5 *If $d < 4$, for every $r > 1$, there exists a $c = c(r) > 0$ such that if $x, y \in \partial C_n$, $z_1, z_2 \in \partial C_{rn}$,*

$$P^{x, y}\{S^1[0, \tau_{rn}^1] \cap S^2[0, \tau_{rn}^2] \ne \emptyset \mid S^1(\tau_{rn}^1) = z_1, S^2(\tau_{rn}^2) = z_2\} \ge c.$$

Proof. Let $\sigma^i = \sigma_{n, r}^i = \tau_{n(r+1)/2}^i$. By Proposition 5.2.4 , the strong Markov property, and Harnack's inequality (Theorem 1.7.2),

$$P^{x, y}\{S^1[0, \tau_{rn}^1] \cap S^2[0, \tau_{rn}^2] \ne \emptyset, S^1(\tau_{rn}^1) = z_1, S^2(\tau_{rn}^2) = z_2\}$$
$$\ge \quad P^{x, y}\{S^1[0, \sigma^1] \cap S^2[0, \sigma^2] \ne \emptyset, S^1(\tau_{rn}^1) = z_1, S^2(\tau_{rn}^2) = z_2\}$$
$$\ge \quad c(r) \inf_{w_1, w_2 \in \partial C_{n(r+1)/2}} P^{w_1, w_2}\{S^1(\tau_{rn}^1) = z_1, S^2(\tau_{rn}^2) = z_2\}$$
$$\ge \quad c(r) P^{x, y}\{S^1(\tau_{rn}^1) = z_1, S^2(\tau_{rn}^2) = z_2\}. \qquad \square$$

Before proceeding with the proof of Theorem 5.2.3 we need one fact about Brownian motion. Let A be any set contained in D with $x \notin A$. Then [60, Theorem 2.6.3]

$$P^x\{P^1[0, T^1] \cap A \text{ is non-empty, finite}\} = 0.$$

If we condition on B^2 and use this fact we get

$$P^{x,y}\{B^1[0, T^1] \cap B^2[0, T^2] \text{ is non-empty, finite}\} = 0. \qquad (5.6)$$

For each $\epsilon > 0$, let $\mathcal{A}_\epsilon$ be the set of all open balls of radius 3ϵ centered at points $(j_1\epsilon, j_2\epsilon, j_3\epsilon)$, where j_1, j_2, j_3 are integers. Note that if $|z - w| < \epsilon$, then there exists an $A \in \mathcal{A}$ with $z, w \in A$. Let X_ϵ be the cardinality of

$$\{A \in \mathcal{A}_\epsilon : B^1[0, T^1] \cap A \neq \emptyset, B^2[0, T^2] \cap A \neq \emptyset\}.$$

If $B^1[0, T^1] \cap B^2[0, T^2] = \emptyset$, then for all sufficiently small ϵ, $X_\epsilon = 0$. Conversely, if $B^1[0, T^1] \cap B^2[0, T^2] \neq \emptyset$, then $X_\epsilon \neq 0$. Therefore,

$$b(x, y, 1) = \lim_{\epsilon \to 0} P\{X_\epsilon = 0\}.$$

If for some $1 \leq k < \infty$ and some $\epsilon_n \to 0$,

$$\lim_{n \to \infty} P\{X_{\epsilon_n} = k\} > 0,$$

then it is easy to see that with positive probability $B^1[0, T^1] \cap B^2[0, T^2]$ contains exactly k points. Since this is impossible by (5.6), for every $k > 0$,

$$b(x, y, 1) = \lim_{\epsilon \to 0} P\{X_\epsilon \leq k\}.$$

For each n, ϵ let $\mathcal{H}_{n,\epsilon}$ be the collection of $C_{4n\epsilon}(z)$ where

$$z = ([nj_1\epsilon], [nj_2\epsilon], [nj_3\epsilon]),$$

with j_1, j_2, j_3 integers, and let $Y_{n,\epsilon}$ be the cardinality of

$$\{H \in \mathcal{H}_{n,\epsilon} : S^1[0, \tau_n^1] \cap H \neq \emptyset, S^2[0, \tau_n^2] \cap H \neq \emptyset\}.$$

By the invariance principle, for each $K < \infty$,

$$\lim_{\epsilon \to 0} \liminf_{n \to \infty} P\{Y_{n,\epsilon} > K\} \geq \lim_{\epsilon \to 0} P\{X_\epsilon > K\} = 1 - b(x, y, 1). \qquad (5.7)$$

Assume $Y_{n,\epsilon} > K$. Then one can find at least $J = J(K) = [(17)^{-3}K]$ sets $H \in \mathcal{H}_{n,\epsilon}$, $C_{4n\epsilon}(z_1), \dots, C_{4n\epsilon}(z_J)$, with $|z_i - z_j| \geq 8n\epsilon$. Let

$$
\begin{aligned}
\eta^i(j) &= \inf\{k : S^i(k) \in \partial C_{4n\epsilon}(z_j)\}, \\
\sigma^i(j) &= \sup\{k \leq \eta^i(j) : S^i(k) \in \partial C_{8n\epsilon}(z_j)\}, \\
\rho^i(j) &= \inf\{k \geq \eta^i(j) : S^i(k) \in \partial C_{8n\epsilon}(z_j)\}.
\end{aligned}
$$

Then the paths $S^i[\sigma^i(j), \rho^i(j)], i = 1, 2, j = 1, \ldots, J$ are conditionally independent given $S^i(\sigma^i(j)), S^i(\rho^i(j)), i = 1, 2, j = 1, \ldots, J$. From Proposition 5.2.5 we conclude that

$$P^{x_n, y_n}\{S^1[0, \tau_n^1] \cap S^2[0, \tau_n^2] \neq \emptyset \mid Y_{n,\epsilon} \geq K\} \geq 1 - (1 - c)^{J(K)}.$$

Since we can make J arbitrarily large, this implies by (5.7) that

$$\liminf_{n \to \infty} P^{x_n, y_n}\{S^1[0, \tau_n^1] \cap S^2[0, \tau_n^2] \neq \emptyset\} \geq 1 - b(x, y, 1).$$

This combined with (5.5) gives Theorem 5.2.3 .

By the same proof we can derive the following.

Corollary 5.2.6 *If $r < 1$, then*

$$\lim_{n \to \infty} P\{S^1[\tau_{rn}^1, \tau_n^1] \cap S^2[\tau_{rn}^2, \tau_n^2] = \emptyset\} = P\{B^1[T_r^1, T_1^1] \cap B^2[T_r^2, T_1^2] = \emptyset\}.$$

5.3   Equivalence of Exponents

In this section we will show that $\xi = 2\zeta$. More precisely, we prove the following.

Theorem 5.3.1 *If $d = 2, 3$, as $n \to \infty$*

$$f(n) \approx n^{-\zeta},$$

where $\zeta = \xi/2$.

By Lemma 1.5.1, for every $\epsilon > 0$, there is an $\alpha = \alpha(\epsilon) > 0$ such that

$$P\{\tau_n^1 \leq n^{2-\epsilon}\} \leq O(\exp\{-n^\alpha\}).$$

Also,

$$
\begin{aligned}
P\{\tau_n^i \geq n^{2+\epsilon}\} &\leq P\{\inf_{0 \leq k < n^\epsilon} |S((k+1)n^2) - S(kn^2)| \leq 2n\} \\
&\leq (1 - c)^{n^\epsilon} \\
&\leq O(\exp\{-n^\alpha\}),
\end{aligned}
$$

for some $\alpha = \alpha(\epsilon)$. Therefore if

$$h(n) = P\{S^1[0, \tau_n^1] \cap S^2[0, \tau_n^2] = \emptyset\},$$

we have for every $\epsilon > 0$,

$$f(n^{2+\epsilon}) \leq h(n) + O(\exp\{-n^\alpha\}) \leq f(n^{2-\epsilon}).$$

Theorem 5.3.1 is therefore an immediate corollary of the following theorem which we wil prove.

Theorem 5.3.2 *If $d = 2, 3$, then as $n \to \infty$,*

$$h(n) \approx n^{-\xi}.$$

We have done most of the work for the lower bound already. Let $\epsilon > 0$ and find $r > \epsilon^{-1}$ such that

$$b(r) = \sup_{|x|,|y|=1} b(x, y, r) \leq r^{\epsilon - \xi}. \tag{5.8}$$

By the strong Markov property,

$$P\{S^1(0, \tau_{rn}^1] \cap S^2(0, \tau_{rn}^2] = \emptyset \mid S^1(0, \tau_n^1] \cap S^2(0, \tau_n^2] = \emptyset\}$$

$$\leq \sup_{x,y \in \partial C_n} P\{S^1(0, \tau_{rn}^1] \cap S^2[0, \tau_{rn}^2] = \emptyset\}.$$

By Theorem 5.2.3 and (5.8), the right hand side is bounded above for sufficiently large n by $2r^{\epsilon - \xi}$. By iterating, for some m sufficiently large, and all $k > 0$,

$$h(r^k m) \leq (2r^{\epsilon - \xi})^k,$$

and hence

$$\liminf_{k \to \infty} -\frac{\ln h(r^k m)}{\ln r^k m} \geq -\frac{\ln 2 + (\epsilon - \xi) \ln r}{\ln r}.$$

Since h is a decreasing function this clearly implies that

$$\liminf_{n \to \infty} -\frac{\ln h(n)}{\ln n} \geq -\frac{\ln 2 + (\epsilon - \xi) \ln r}{\ln r},$$

Since for every $\epsilon > 0$ this holds for all r sufficiently large,

$$\liminf_{n \to \infty} -\frac{\ln h(n)}{\ln n} \geq \xi.$$

The upper bound is proved by finding an appropriate Brownian motion event on which the invariance principle can be applied. Suppose two Brownian motion paths do not intersect. Then one would expect that they would stay a "reasonable distance apart"—at least as far apart as unconditioned Brownian motions do. It can be very difficult to prove such facts. The next proposition asserts a weak form of this intuitive fact. Even this relatively weak statement has a fairly technical proof using ideas of excursions of Brownian motions. Rather than develop the necessary machinery to discuss the proof, which is not very illuminating, we will omit it. See [11] for details.

Lemma 5.3.3 *Let $A_\epsilon(r)$ be the event*

$$(i) \quad B^1[0, T_r^1] \cap B^2[0, T_r^2] = \emptyset,$$
$$(ii) \quad \mathrm{dist}(B^i(T_r^i), B^{3-i}[0, T_r^{3-i}]) \geq \epsilon r, \quad i = 1, 2,$$
$$(iii) \quad B^i[0, T_r^i] \cap D \subset D_{\epsilon/8}(B^i(0)), \quad i = 1, 2.$$

Let

$$v(\epsilon, r) = \inf\{P^{x,y}(A_\epsilon(r)) : x, y \in \partial D, |x - y| \geq \frac{\epsilon}{4}\}.$$

Then for some $\epsilon > 0$,

$$\liminf_{r \to \infty} -\frac{\ln v(\epsilon, r)}{\ln r} = \xi.$$

To prove the upper bound from the lemma, let $\xi_0 > \xi$ and $\epsilon > 0$ be as in Lemma 5.3.3. Let $U(n, r)$ be the event

$$(i) \quad S^1[0, \tau_{rn}] \cap S^2[0, \tau_{rn}^2] = \emptyset$$
$$(ii) \quad \mathrm{dist}(S^i(\tau_{rn}^i), S^{3-i}[0, \tau_{rn}^{3-i}]) \geq \epsilon n/2, \quad i = 1, 2$$
$$(iii) \quad S^i[0, \tau_{rn}^i] \cap C_n \subset C_{\epsilon n/4}(S^i(0)), \quad i = 1, 2.$$

By Lemma 5.3.3, there exist arbitrarily large r with $v(\epsilon, r) \geq r^{-\xi_0}$. For such an r, by the invariance principle, there exists an $m = m(r, \xi_0)$ such that for $n \geq m$, $x, y \in C_n$, $|x - y| \geq \epsilon n/2$,

$$P^{x,y}(U(n, r)) \geq \frac{1}{2} r^{-\xi_0}.$$

By iterating and using the strong Markov property, one can easily check for some $c = c(m)$,

$$h(r^k m) \geq c2^{-k} r^{-k\xi_0}.$$

Therefore,

$$\limsup_{k \to \infty} -\frac{\ln h(r^k m)}{\ln(r^k m)} \leq \frac{\xi_0 \ln r + \ln 2}{\ln r},$$

and hence by the monotonicity of h,

$$\limsup_{n \to \infty} -\frac{\ln h(n)}{\ln n} \leq \frac{\xi_0 \ln r + \ln 2}{\ln r}.$$

Since for any $\xi_0 > \xi$ this holds for a sequence $r_j \to \infty$,

$$\limsup_{n \to \infty} -\frac{\ln h(n)}{\ln n} \leq \xi. \quad \square$$

The same ideas apply to intersections of more than two walks. Let $B^1, B^2, \ldots, B^{k+1}$ be independent Brownian motions starting at $e_1, -e_1, -e_1, \ldots, -e_1$ respectively and let

$$q(r,k) = P^{e_1,-e_1,\ldots,-e_1}\{B^1[0,T_r^1] \cap (B^2[0,T_r^2] \cup \cdots \cup B^{k+1}[0,T_r^{k+1}]) = \emptyset\}.$$

Similarly, if $S^1, S^2, \ldots, S^{k+1}$ are independent simple random walks starting at 0 we let (as in section 3.7)

$$f(n,k) = P\{S^1(0,n] \cap (S^2(0,n] \cup \cdots \cup S^{k+1}(0,n]) = \emptyset\}.$$

Then the following theorem can be proved in the same fashion as above.

Theorem 5.3.4 *If $d = 2, 3$, there exist $\xi(k) = \xi_d(k)$ such that as $r \to \infty$,*

$$q(r,k) \approx r^{-\xi(k)}.$$

Moreover, as $n \to \infty$,
$$f(n,k) \approx n^{-\zeta(k)},$$

where $\zeta(k) = \xi(k)/2$.

We can restate Theorem 3.5.1 by

$$\zeta_2(2) = 1, \quad \zeta_3(2) = \frac{1}{2},$$

and hence

$$\xi_2(2) = 2, \quad \xi_3(2) = 1. \tag{5.9}$$

5.4 Variational Formulation

Assume $B^1(0) = e_1$ and let Q_r be the conditional probability of $\{B^1[0,T_r^1] \cap B^2[0,T_r^2] = \emptyset\}$ given B^1, i.e.,

$$Q_r(\omega_1) = P_2^{-e_1}\{B^1[0,T_r^1,\omega_1] \cap B^2[0,T_r^2,\omega_2] = \emptyset\}.$$

Then,

$$q(r,k) = E_1(Q_r^k).$$

Also, by integration by parts,

$$q(r,k) = \int_0^1 x^k dP\{Q_r \leq x\} = \int_0^1 kx^{k-1} P\{Q_r \geq x\} dx. \tag{5.10}$$

In particular, for any $x \in [0, 1]$,

$$q(r, k) \geq x^k P\{Q_r \geq x\}. \tag{5.11}$$

For any $a > 0$ we define

$$\beta(a) = \beta_d(a) = \liminf_{r \to \infty} -\frac{\ln P\{Q_r \geq r^{-a}\}}{\ln r}.$$

One can actually show that the limit exists; however, this requires a lot of technical work and will be unnecessary for what we will discuss below. Intuitively we think of $\beta(a)$ by

$$P\{Q_r \geq r^{-a}\} \approx r^{-\beta(a)}.$$

The next proposition is a standard kind of result from the subject of large deviations [14, 21] where such variational formulations go under the name Legendre transforms.

Proposition 5.4.1

$$\xi(k) = \inf_{a>0}(ak + \beta(a)). \tag{5.12}$$

Proof. For any $a > 0, \delta > 0$ there is a sequence $r_n \to \infty$ with

$$P\{Q_{r_n} \geq r_n^{-a}\} \geq r_n^{-(\beta(a)+\delta)}.$$

Therefore by (5.11),

$$q(r_n, k) \geq r_n^{-ak-\beta(a)-\delta},$$

and hence

$$\liminf_{r \to \infty} -\frac{\ln q(r, k)}{\ln r} \leq ak + \beta(a) + \delta.$$

By Theorem 5.3.4, this implies

$$\xi(k) \leq ak + \beta(a) + \delta.$$

Since this holds for every $a, \delta > 0$,

$$\xi(k) \leq \inf_{a>0}(ak + \beta(a)).$$

To prove the other direction, let

$$q = q(k) = \inf_{a>0}(ak + \beta(a)).$$

For any $0 < \delta < \frac{1}{2}$, let $M = M(\delta)$ be an integer satisfying

$$\delta^{-1}\xi(k) \leq M \leq 2\delta^{-1}\xi(k).$$

For any $\gamma > 0$, for all r sufficiently large,

$$P\{Q_r \geq r^{-j\delta}\} \leq r^{-\beta(j\delta)+\gamma}, j = 1, \ldots, M. \tag{5.13}$$

By (5.10),

$$
\begin{aligned}
q(r, k) &= \int_0^1 kx^{k-1}P\{Q_r \geq x\}dx \\
&\leq kr^{-M} + k\sum_{j=1}^M \int_{r^{-j\delta}}^{r^{-(j-1)\delta}} x^{k-1}P\{Q_r \geq x\}dx.
\end{aligned}
$$

By (5.13), for r sufficiently large,

$$\int_{r^{-j\delta}}^{r^{-(j-1)\delta}} x^{k-1}P\{Q_r \geq x\}dx \leq r^{-(j-1)(k-1)\delta}r^{-\beta(j\delta)+\gamma}(r^{-(j-1)\delta} - r^{-j\delta}).$$

Therefore for r sufficiently large,

$$
\begin{aligned}
\sum_{j=1}^M \int_{r^{-j\delta}}^{r^{-(j-1)\delta}} x^{k-1}P\{Q_r \geq x\}dx &\leq r^{\gamma+\delta(k-1)}\sum_{j=1}^M r^{-k\delta j-\beta(j\delta)}(r^\delta - 1) \\
&\leq r^{\gamma+\delta k}Mr^{-q} \\
&\leq 2\delta^{-1}r^{\gamma+\delta k-q}\xi(k),
\end{aligned}
$$

and,

$$q(r, k) \leq kr^{-\xi(k)/\delta} + 2\delta^{-1}kr^{\gamma+\delta k-q}\xi(k).$$

For δ sufficiently small the second term dominates the first term and we get

$$\xi(k) = \liminf_{r\to\infty} -\frac{\ln q(r, k)}{\ln r} \geq q - \gamma - \delta k.$$

Since this is true for all $\delta, \gamma > 0$, we have $\xi(k) \geq q$. $\square$

Our estimates in the remainder of this chapter will be derived by getting bounds on $\beta(a)$; using the value for $\xi(2)$ (see (5.9)); and using Proposition 5.4.1. As an example of how this proposition can be used we will derive Hölder's inequality. Suppose $j < k$. Then

$$
\begin{aligned}
\xi(k) &= \inf_{a>0}(ka + \beta(a)) \\
&\leq \frac{k}{j}\inf_{a>0}(ja + \beta(a)) = \frac{k}{j}\xi(j),
\end{aligned}
$$

i.e.,

$$\frac{j}{k}\xi(k) \leq \xi(j) \leq \xi(k).$$

5.5   Lower Bound in Two Dimensions

If B^1 is a Brownian motion in R^2 and $r > 0$, let V_r be the event

$$\{B^1[0, T_r^1] \text{ disconnects } 0 \text{ and } \partial D_r\}.$$

More precisely, V_r is the event that the connected component of $R^2 \setminus B^1[0, T_r^1]$ containing 0 is contained in the open ball D_r. It is easy to check that if $0 < |x| < r$,

$$0 < P^x(V_r) < 1.$$

If $|e| = 1$, let

$$\psi(r) = 1 - P^e(V_r) = P^e(V_r^c).$$

For general $0 < |x| < r$, we have by scaling

$$P^x(V_r^c) = \psi(\frac{r}{|x|}).$$

Also, if $1 < r, s, < \infty$, $|e| = 1$,

$$\psi(rs) = P^e(V_{rs}^c) \le P^e(V_r^c)P^{B^1(T_r)}(V_{rs}^c) = \psi(r)\psi(s). \tag{5.14}$$

Therefore, $\phi(k) = \ln \psi(2^k)$ is a subadditive function, and the following proposition is an immediate corollary of Lemma 5.2.1 .

Proposition 5.5.1 *If $d = 2$,*

$$\lim_{r\to\infty} -\frac{\ln \psi(r)}{\ln r} = \sup_{r>1} -\frac{\ln \psi(r)}{\ln r} \doteq \overline{\gamma} > 0.$$

We can now derive a lower bound for $\xi = \xi(1)$ in terms of $\overline{\gamma}$. Suppose B^1 starts at e_1. Then on the event V_r, $Q_r = 0$, where Q_r is as defined in the last section. Hence for any $a > 0$,

$$P\{Q_r \ge r^{-a}\} \le P^{e_1}(V_r^c) = \psi(r),$$

and hence by Proposition 5.5.1, for all $a > 0$,

$$\beta(a) \ge \overline{\gamma}. \tag{5.15}$$

Proposition 5.5.2 *If $d = 2$,*

$$\xi = \xi(1) \ge 1 + \frac{1}{2}\overline{\gamma}.$$

Proof. By (5.9) and (5.12), for every $a > 0$,

$$2 = \xi(2) \leq 2a + \beta(a),$$

and hence $a \geq 1 - \frac{1}{2}\beta(a)$. Therefore, by (5.12) and (5.15),

$$\xi(1) = \inf_{a>0}(a + \beta(a)) \geq \inf_{a>0}(1 + \frac{1}{2}\beta(a)) \geq 1 + \frac{1}{2}\overline{\gamma}. \quad \square$$

Bounds on $\overline{\gamma}$ will thus produce bounds on ξ. The remainder of this section will be devoted to proving the following bound on $\overline{\gamma}$.

Proposition 5.5.3

$$\overline{\gamma} \geq \pi^{-2}.$$

This is not the optimal result. By similar, but technically more complicated means, one can prove that $\overline{\gamma} \geq (2\pi)^{-1}$. Even this bound is probably not correct—it has been conjectured that $\overline{\gamma} = 1/4$. To prove Proposition 5.5.3, we will use conformal invariance properties of Brownian motion in R^2. We may consider a two-dimensional Brownian motion B_t as a complex-valued Brownian motion

$$B_t = \Re(B_t) + i\Im(B_t),$$

where $\Re(B_t)$ and $\Im(B_t)$ are independent one-dimensional Brownian motions. Suppose $h : C \to C$ is an analytic function and

$$v_t = \int_0^t |h'(B_s)| ds,$$

$$\gamma_t = \inf\{s : v_s \geq t\}.$$

Then [18]

$$\gamma_t = h(B(v_t))$$

is also a complex-valued Brownian motion. Roughly speaking, an analytic function transforms a Brownian motion into a time change of a Brownian motion.

Proof of Proposition 5.5.3. Let $|e| = 1$ and

$$\overline{\psi}(r) = \psi(\frac{1}{r}) = P^{er}(V_1^c).$$

Then by Proposition 5.5.1,

$$\overline{\gamma} = \sup_{0<r<1} \frac{\ln \overline{\psi}(r)}{\ln r}. \tag{5.16}$$

Let B_t be a complex-valued Brownian motion starting at x, where $x < 0$, and let $h(z) = \exp(z)$. Let

$$\tau_0 = \inf\{t : \Re(B_t) = 0\}.$$

Then $Y_t = \exp(B_t), 0 \le t \le \tau_0$ is (a time change of) a Brownian motion starting at $\exp(x)$, stopped when it reaches ∂D. Therefore, the distribution of the random set

$$\{\exp(B_t) : 0 \le t \le \tau_0 \mid B_0 = x\}$$

is the same as the distribution of

$$\{B_t : 0 \le t \le T^1 \mid B_0 = \exp(x)\}.$$

Let $r \in (0,1)$, $x = \ln r$ and assume $B_0 = x$. If $w < x$, let

$$\tau_w = \inf\{t : \Re(B_t) = w\},$$

$$\sigma = \sup\{t < \tau_w : \Re(B_t) = x\}.$$

$$\eta = \inf\{t > \tau_w : \Re(B_t) = x\}.$$

Suppose that $\tau_w < \tau_0$ and

$$|\Im(B_\eta) - \Im(B_\sigma)| \ge 2\pi.$$

Then it follows that the path

$$\{\exp(B_t) : 0 \le t \le \tau_0\}$$

disconnects 0 and ∂D. Therefore,

$$\psi(r) \le 1 - P^x\{\tau_w < \tau_0, |\Im(B_\eta) - \Im(B_\sigma)| \ge 2\pi\}.$$

Note that by the "gambler's ruin" estimate for one-dimensional Brownian motion,

$$P^x\{\tau_w < \tau_0\} = \frac{x}{w}.$$

To compute the conditional probability, we will need the hitting distribution of the imaginary axis. If $B_0 = x$, then the density of $\Im(B_{\tau_0})$ is given by [18, (1.92)]

$$\frac{|x|}{\pi(x^2 + y^2)}, \qquad -\infty < y < \infty.$$

Therefore,

$$P^x\{|\Im(B_\eta) - \Im(B_\sigma)| \geq 2\pi \mid \tau_w < \tau_0\}$$

$$= P^{B(\tau_w)}\{|\Im(B_{\tau_x}) - \Im(B_\sigma)| \geq 2\pi\}$$

$$\geq 1 - \int_{-2\pi+\mathcal{I}(B_\sigma)}^{2\pi+\mathcal{I}(B_\sigma)} \frac{w - x}{\pi((w - x)^2 + y^2)} dy$$

$$\geq 1 - \int_{-2\pi}^{2\pi} \frac{w - x}{\pi((w - x)^2 + y^2)} dy$$

$$= 1 - \frac{2}{\pi} \arctan(\frac{2\pi}{w - x}).$$

Therefore, for every $w < \ln r < 0$,

$$\psi(r) \leq 1 - \frac{\ln r}{w}[1 - \frac{2}{\pi} \arctan(\frac{2\pi}{w - \ln r})].$$

Since

$$\lim_{y \to 0+} \frac{1}{y}[1 - \frac{2}{\pi} \arctan(\frac{2\pi}{y})] = \pi^{-2},$$

for every $\epsilon > 0$ we can find $w < 0$ and $R \in (e^w, 1)$ such that for all $r \in (R, 1)$,

$$\psi(r) \leq 1 - \pi^{-2}(1 - \epsilon)\ln r,$$

and hence by (5.16),

$$\overline{\gamma} \geq \limsup_{r \to 1-} \frac{\ln \overline{\psi}(r)}{\ln r}$$

$$\geq \limsup_{r \to 1-} \frac{\ln[1 - \pi^{-2}(1 - \epsilon) \ln r]}{\ln r}$$

$$= \pi^{-2}(1 - \epsilon).$$

Since this holds for all $\epsilon > 0$, the proposition is proved. □

5.6 Upper Bound

The main result in this section is the following.

Proposition 5.6.1 *(a) If $d = 2$,*

$$\lim_{a \to \frac{1}{2}+} \beta(a) = \infty.$$

(b) If $d = 3$,

$$\lim_{a \to 0+} \beta(a) = \infty.$$

Corollary 5.6.2 (a) If $d = 2$,

$$\xi = \xi(1) < \frac{3}{2}.$$

(b) If $d = 3$,

$$\xi = \xi(1) < 1.$$

Proof. Assume $d = 2$. By Proposition 5.6.1(a), there exists $a_0 > \frac{1}{2}$ such that $\beta(a) > 2$ for $a \leq a_0$. By (5.9) and (5.12), for any $\epsilon \in (0,1)$ we can find $a = a_\epsilon$ such that

$$2 = \xi(2) \geq 2a + \beta(a) - \epsilon.$$

Clearly $a \geq a_0$. Since $\beta(a) \leq 2 - 2a + \epsilon$, by (5.12),

$$\xi = \xi(1) \leq a + \beta(a) \leq 2 - a + \epsilon \leq 2 - a_0 + \epsilon.$$

Therefore, $\xi \leq 2 - a_0 < \frac{3}{2}$. This gives (a); (b) is proved similarly. $\quad\square$

The proof of Proposition 5.6.1 will need a simple large deviations estimate for binomial random variables. The following lemma can be deduced easily from "Chernoff's theorem" [7, Theorem 9.3].

Lemma 5.6.3 *For every $b < \infty$, there exist $p = p(b) < 1, \delta = \delta(b) > 0, C = C(b) < \infty$, such that if Y is a binomial random variable with parameters n and p,*

$$P\{Y \leq \delta n\} \leq Ce^{-nb}.$$

Proof of Proposition 5.6.1. We first consider $d = 3$. Let Z be the conditional probability of

$$B^1[T^1_{3/2}, T^1_2] \cap B^2[0, T^2_2] = \emptyset,$$

given B^1 (assuming $B^1(0) = e_1, B^2(0) = -e_1$). It is not difficult to show that $P^{e_1}_1\{Z = 0\} = 0$ and hence

$$\lim_{\epsilon \to 0} P^{e_1}_1\{Z \geq \epsilon\} = 1.$$

By the standard Harnack inequality for harmonic functions, this implies

$$\lim_{\epsilon \to 0} P^{e_1}_1\{\inf_{x \in D} P^x_2\{B^1[T^1_{3/2}, T^1_2] \cap B^2[0, T^2_2] \neq \emptyset\} \geq \epsilon\} = 1.$$

For $j = 1, 2, \ldots$, let $J_{j,\epsilon}$ be the indicator function of the event

$$\inf_{x \in D_{2j-1}} P^x_2\{B^1[T^1_{3 \cdot 2^{j-2}}, T^1_{2^j}] \cap B^2[0, T^2_{2^j}] \neq \emptyset\} \geq \epsilon,$$

and let

$$Y_{k,\epsilon} = \sum_{j=1}^{k} J_{j,\epsilon}.$$

Then by the strong Markov property applied to B^2,

$$Q_{2^k} \le (1-\epsilon)^{Y_{k,\epsilon}},$$

and hence

$$
\begin{aligned}
P\{Q_{2^k} \ge 2^{-ka}\} &\le P\{(1-\epsilon)^{Y_{k,\epsilon}} \ge 2^{-ka}\} \\
&= P\{Y_{k,\epsilon} \le -\frac{ka\ln 2}{\ln(1-\epsilon)}\}.
\end{aligned}
$$

The random variable $Y_{k,\epsilon}$ is bounded below by a random variable with a binomial distribution with parameters k and $q(\epsilon)$ where $q(\epsilon) \to 1$ as $\epsilon \to 0$. For any $b < \infty$, find p, δ, C as in Lemma 5.6.3. Choose $\epsilon > 0$ so that $q(\epsilon) \ge p$ and then choose $a > 0$ so that

$$-\frac{a\ln 2}{\ln(1-\epsilon)} \le \delta.$$

Then,

$$P\{Y_{k,\epsilon} \le -\frac{ka\ln 2}{\ln(1-\epsilon)}\} \le Ce^{-kb} = C(2^k)^{-b/\ln 2}.$$

Hence,

$$\beta(a) \ge \frac{b}{\ln 2},$$

and

$$\lim_{a\to 0+} \beta(a) = \infty,$$

which gives Proposition 5.6.1(b) .

The proof of Proposition 5.6.1(a) relies on some results from complex function theory. Let B_t^1, B_t^2 be independent complex-valued Brownian motions starting at 0 and πi respectively stopped at

$$\sigma_r^j = \inf\{t : \Re(B_t^j) = \ln r\}.$$

Then $\tilde{B}_t^1 = \exp(B_t^1), \tilde{B}_t^2 = \exp(B_t^2)$ are independent (time changes of) Brownian motions starting at $e_1, -e_1$ respectively stopped at

$$T_r^j = \inf\{t : |\tilde{B}_t^j| = r\}.$$

Let Γ_r be the random set

$$\Gamma_r = \{B_t^2 + 2\pi i k : 0 \le t \le \sigma_r^2, k \in Z\}.$$

Then $\tilde{B}^1[0, T_r^1] \cap \tilde{B}^2[0, T_r^2] = \emptyset$ if and only if

$$B^1[0, \sigma_r^1] \cap \Gamma_r = \emptyset.$$

For fixed Γ_r,

$$P_1\{B^1[0, \sigma_r^1] \cap \Gamma_r = \emptyset\} = \omega(0, C \setminus (\Gamma_r \cup V_r), V_r),$$

where $V_r = \{\Re(z) = \ln r\}$ and ω denotes harmonic measure in R^2 (here we are using the notation of [1, Chapter 4]). Therefore, if we define Q_r as in section 4 using the Brownian motions $\tilde{B}_t^1, \tilde{B}_t^2$, we get

$$Q_r = \omega(0, C \setminus (\Gamma_r \cup V_r), V_r).$$

Note that Γ_r is a continuous curve connecting πi to V_r along with all the $2\pi i$ "translates" of the curve. It is a result of Beurling that the harmonic measure of V_r is maximized if we take Γ_r to be a straight line parallel to the real axis. The next lemma estimates the harmonic measure in this case. Let $B_t = B_t^1$.

Lemma 5.6.4 *Let*

$$\begin{aligned}
\Lambda_r &= \{z \in C : 0 \le \Re(z) \le \ln r, \Im(z) = (2k+1)\pi i, k \in Z\}, \\
\sigma_r &= \inf\{t : \Re(B_t) = \ln r\}, \\
\tau_r &= \inf\{t : B_t \in \Lambda_r\}.
\end{aligned}$$

Then there exists a constant $c < \infty$ such that for all z with $\Re(z) \le 0$,

$$P^z\{\sigma_r < \tau_r\} = \omega(z, C \setminus (\Lambda \cup V_r), V_r) \le cr^{-1/2}.$$

Proof. Consider

$$\begin{aligned}
\overline{\Lambda}_r &= \{z \in C : -\infty < \Re(z) \le \ln r, \Im(z) = (2k+1)\pi i, k \in Z\}, \\
\overline{\tau}_r &= \inf\{t : B_t \in \overline{\Lambda}_r\}.
\end{aligned}$$

In this case the harmonic measure of V_r can be computed exactly for $\Re(z) \le \ln r$ by recalling that the harmonic measure of V_r is the harmonic function on $C \setminus (\overline{\Gamma}_r \cup V_r)$ with boundary values 1 on V_r and 0 on $\overline{\Lambda}_r$. The solution using separation of variables (see, e.g., [4, 11.9.8]) is

$$\begin{aligned}
P^{x+iy}\{\tau < \overline{\tau}_r\} &= \sum_{n=0}^{\infty} \frac{4}{(2n+1)\pi} e^{-(n+\frac{1}{2})(\ln r - x)} \cos[(n+\frac{1}{2})y] \\
&\le cr^{-1/2}, \text{ if } x \le 0.
\end{aligned} \qquad (5.17)$$

Now assume $\ln r \geq \pi$ (it clearly suffices to prove the lemma for sufficiently large r). By symmetry of Brownian motion, it is not difficult to show in this case that if $\Re(z) = 0$,

$$P^z\{\overline{\tau}_r = \tau_r\} \geq \frac{1}{8}. \tag{5.18}$$

Let

$$g(r) = \sup_{\Re(z) \leq 0} P^z\{\sigma_r < \tau_r\} = \sup_{\Re(z) = 0} P^z\{\sigma_r < \tau_r\}.$$

If $\Re(z) = 0$,

$$P^z\{\sigma_r < \tau_r\} = P^z\{\sigma_r < \overline{\tau}_r\} + P^z\{\sigma_r < \tau_r, \overline{\tau}_r < \sigma_r \wedge \tau_r\}.$$

By (5.18), $P^z\{\overline{\tau}_r < \sigma_r \wedge \tau_r\} \leq \frac{7}{8}$ and by the strong Markov property,

$$P^z\{\sigma_r < \tau_r \mid \overline{\tau}_r < \sigma_r \wedge \tau_r\} \leq g(r).$$

Therefore by (5.17),

$$g(r) \leq cr^{-1/2} + \frac{7}{8}g(r),$$

or

$$g(r) \leq cr^{-1/2}. \quad \square$$

Let $\gamma : [0, T] \to C$ be any continuous path with $\Re(\gamma(0)) = 0, \Re(\gamma(T)) = \ln r$, and

$$0 < \Re(\gamma(t)) < \ln r, 0 < t < T.$$

Let $\Gamma = \Gamma_\gamma$ be the corresponding set of translates by $2\pi i$,

$$\Gamma = \{\gamma(t) + 2\pi i k : 0 \leq t \leq T, k \in Z\}.$$

Let $\overline{\Gamma}$ be the extension by straight lines on $\{\Re(z) \leq 0\}$, i.e.,

$$\overline{\Gamma} = \Gamma \cup \{\gamma(0) + 2\pi i k - s : k \in Z, s \geq 0\}.$$

By an argument similar to that in Lemma 5.6.4,

$$\omega(0, C \setminus (\Gamma \cup V_r), V_r) \leq 8\omega(0, C \setminus (\overline{\Gamma} \cup V_r), V_r). \tag{5.19}$$

If $0 \in \overline{\Gamma}$, then $\omega(0, C \setminus (\overline{\Gamma} \cup V_r), V_r) = 0$. Assume $0 \notin \overline{\Gamma}$ and let U be the (open) connected component of $C \setminus (\overline{\Gamma} \cup V_r)$ containing 0. If $\partial U \cap V_r = \emptyset$, then $\omega(0, C \setminus (\overline{\Gamma} \cup V_r), V_r) = 0$, so assume $\partial U \cup V_r = \emptyset$. Then $\partial U \cap V_r$ is a closed interval of length 2π. We can find a conformal map h taking U onto $W = \{z : \Re(z) < 0, -\pi i < \Im(z) < \pi i\}$ such that $\partial U \cup V_r$ is mapped onto $V = \{\Re(z) = 0, -\pi i \leq \Im(z) \leq \pi i\}$. By the conformal invariance of harmonic measure (or equivalently the conformal invariance of Brownian motion),

$$\omega(0, C \setminus (\overline{\Gamma} \cup V_r), V_r) = \omega(h(0), W, V). \qquad (5.20)$$

Note that by Lemma 5.6.4,

$$\omega(h(0), W, V) \leq c e^{\Re(h(0))/2}. \qquad (5.21)$$

For each $0 \leq x \leq \ln r$, let $\theta(x)$ be the length of $U \cap \{\Re(z) = x\}$. By (4-23) of [1], the map h must satisfy

$$\Re(h(0)) \leq -2\pi \int_0^{\ln r} \frac{dx}{\theta(x)} + 2 \ln 32,$$

and hence by (5.19) - (5.21),

$$\omega(0, C \setminus (\Gamma \cup V_r), V_r) \leq c \exp\{-\pi \int_0^{\ln r} \frac{dx}{\theta(x)}\}. \qquad (5.22)$$

Note that $\theta(x) \leq 2\pi$ so this gives the result

$$\omega(0, C(\Gamma \cup V_r), V_r) \leq c r^{-1/2}.$$

We now apply the inequality (5.22) to the Brownian motion $B_t = B_t^2$ stopped at $\sigma_r = \sigma_r^2$. The Brownian motion path does not satisfy the condition $0 < \Re(B_t) < \ln r$, $t \in (0, \sigma_r)$; however, we can consider instead the "excursion" $B[\eta_r, \sigma_r]$ where

$$\eta_r = \sup\{t < \sigma_r : \Re(B_t) = 0\}.$$

For any $\epsilon > 0$ let $q(\epsilon)$ be the probability that a Brownian motion starting at 0, stopped at $\rho = \inf\{t : |\Re(B_t)| = \frac{1}{2}\}$ encloses the circle of radius ϵ around 0, i.e., the probability that the boundary of the connected component of $C \setminus B[0, \rho]$ does not intersect the closed ball of radius ϵ. Note (see section 5) that

$$\lim_{\epsilon \to 0} q(\epsilon) = 1.$$

Also, by symmetry, the conditional probability of this event given $\{\Re(B_\rho) = \frac{1}{2}\}$ (or given $\{\Re(B_\rho) = -\frac{1}{2}\}$) is still $q(\epsilon)$.

For any j, let

$$\tau_j = \tau_{j,r} = \inf\{t > \eta_r : \Re(B_t) = j + \frac{1}{2}\},$$

$$\rho_j = \rho_{j,r} = \inf\{t \geq \tau_j : \Re(B_t) \in \{j, j+1\}\},$$

and let $I(j, \epsilon)$ be the indicator function of the event that $B[\tau_j, \rho_j]$ encloses the circle of radius ϵ around $B(\tau_j)$. Note that $E(I(j, \epsilon)) = q(\epsilon)$ and that $\{I(j, \epsilon) : j = 0, 1, \ldots [\ln r] - 1\}$ are independent random variables. (This is not immediately obvious since η_r is not a stopping time and hence neither are τ_j or ρ_j. However, τ_j and ρ_j are stopping times for the Brownian "excursion", and the last sentence of the previous paragraph shows that this conditioning does not affect the probability of this event.) If $I(j, \epsilon) = 1$, the path $B[\eta_r, \sigma_r]$ satisfies

$$\theta(x) \leq 2\pi - \frac{\epsilon}{2}, \quad j + \frac{1}{2} - \frac{\epsilon}{2} \leq x \leq j + \frac{1}{2} + \frac{\epsilon}{2},$$

and hence

$$2\pi \int_j^{j+1} \frac{dx}{\theta(x)} \geq 1 + \frac{\epsilon^2}{4\pi}.$$

Therefore,

$$2\pi \int_0^{\ln r} \frac{dx}{\theta(x)} \geq \ln r + \left[\sum_{j=0}^{[\ln r]-1} I(j, \epsilon) \right] (\frac{\epsilon^2}{4\pi}).$$

For every $b < \infty$, find p, δ, C as in Lemma 5.6.3. Choose $\epsilon > 0$ so that $q(\epsilon) \geq p$. Then

$$P\{ \sum_{j=0}^{[\ln r]-1} I(j, \epsilon) \leq \delta[\ln r]\} \leq cr^{-b},$$

and hence

$$P\{2\pi \int_0^{\ln r} \frac{dx}{\theta(x)} \leq [\ln r](1 + \frac{\delta\epsilon^2}{4\pi})\} \leq cr^{-b},$$

and hence by (5.22),

$$P\{\omega(0, C \setminus B[\eta_r, \sigma_r], V_r) \geq cr^{-a}\} \leq cr^{-b},$$

where $a = \frac{1}{2}(1 + \frac{\delta\epsilon^2}{4\pi})$. Therefore,

$$\lim_{a \to \frac{1}{2}+} \beta(a) = \infty. \quad \square$$

Chapter 6

Self-Avoiding Walks

6.1 Introduction

The study of self-avoiding walks arose in chemical physics as a model for long polymer chains. Roughly speaking, a polymer is composed of a large number of monomers which can form together randomly except that the monomers cannot overlap. This restriction is modelled by a self-repulsion term.

The simplest mathematical model to state with such a self-repulsion term is the self-avoiding walk (SAW). A self-avoiding walk of length n is a simple random walk path which visits no site more than once. This simple model does seem to possess many of the qualitative features of polymers. However, it turns out that it it extremely difficult to obtain rigorous results about SAW's, especially in low dimensions which are the most interesting from a physical point of view.

The next two sections discuss the SAW problem. The most interesting characteristics of the model are the dimension dependent critical exponents discussed in Section 6.3. The discussion there is entirely heuristic and mathematicians are still a long way from making the discussion rigorous. One major result [10, 62, 63] is a proof that in high dimensions the exponents take on "mean-field" values. The proof of this result has a field-theoretic flavor of mathematical physics and makes use of a technical tool called the "lace expansion", which has since been applied to some other models in mathematical physics. Because the proof is long and the methods of the proof are significantly different than those discussed in this book, we will not discuss the proof.

There are a number of other ways to put self-repulsion terms on random walks. These split naturally into two categories: configurational measures where random walks are weighted by the number of self-intersections (the original SAW problem is of this type) and kinetically growing measures where random walks are produced from some (non-Markovian) transition functions. It turns out that different self-repulsion terms can give qualitatively different behavior of random walks. We discuss some of these models in Sections 6.4 and 6.5. In the final section we discuss briefly some algorithms used in Monte Carlo simulations of SAW's.

6.2　Connective Constant

A *self-avoiding walk (SAW)* of length n, $\omega = [\omega(0), \ldots, \omega(n)]$ is an ordered sequence of points in Z^d with $|\omega(i) - \omega(i-1)| = 1, i = 1, \ldots, n$ and $\omega(i) \neq \omega(j), 0 \leq i < j \leq n$. In other words, a SAW is a simple random walk path which visits no point more then once. We let Γ_n be the set of SAW's starting at 0 (i.e., $\omega(0) = 0$) and Λ_n be the set of simple random walk paths starting at 0. Note that $|\Lambda_n| = (2d)^n$ and $\Gamma_n \subset \Lambda_n$.

The first question to ask is how many SAW's are there? Let $C_n = |\Gamma_n|$. Since a SAW cannot return to the point it most recently visited, $C_n \leq (2d)(2d-1)^{n-1}$. However, any simple random walk which takes only positive steps in each component is clearly self-avoiding. Since there are d choices at each step for such walks,

$$d^n \leq C_n \leq (2d)(2d-1)^{n-1}. \tag{6.1}$$

Proposition 6.2.1 *There exists a $\mu = \mu_d \in [d, 2d-1]$ such that*

$$C_n \approx \mu^n.$$

Proof. Any $(n+m)$-step SAW consists of an n-step SAW and an m step SAW (although not every choice of an n-step SAW and an m-step SAW can be put together to form an $(n+m)$-step SAW). Therefore,

$$C_{n+m} \leq C_n C_m,$$

and $\phi(n) = \ln C_n$ is a subadditive function. By Lemma 5.2.1,

$$\lim_{n \to \infty} \frac{\phi(n)}{n} = \inf \frac{\phi(n)}{n} \doteq a.$$

Therefore $C_n \approx \mu^n$ where $\mu = e^a$. From (6.1) we get that $\mu \in [d, 2d-1]$. □

The exact value of μ, which is called the *connective constant*, is not known. For $d = 2$, μ is expected to be about 2.64 and μ_3 is expected to be about 4.68. It is rigorously known that $\mu_2 \in (2.58, 2.73)$ [6, 71]. In principle one can calculate μ to any accuracy by finite calculations, but the convergence rate is very slow. As d gets large, the main effect of the self-avoidance constraint is to forbid immediate reversals; Kesten [34] proved that as $d \to \infty$,

$$\mu_d = (2d - 1) - \frac{1}{2d} + O(\frac{1}{d^2}).$$

Kesten [33] also proved that $C_{n+2}/C_n \to \mu^2$, but the conjecture $C_{n+1}/C_n \to \mu$ is still open.

A *self-avoiding polygon (SAP)* of length n is an ordered sequence of points $\omega \in [\omega(0), \ldots, \omega(n)]$ with $|\omega(i) - \omega(i-1)| = 1, 1 \le i \le n; \omega(i) \ne \omega(j), 0 \le i < j \le n - 1$, and $\omega(0) = \omega(n)$. Loosely speaking, a SAP is a self-avoiding walk conditioned to return to its starting point. Let A_n be the number of SAP's of length n. It is easy to see that

$$A_n = \sum_{|e|=1} C_{n-1}(e) = 2dC_{n-1}(e_1),$$

where

$$C_j(x) = |\{\omega \in \Gamma_j : \omega(j) = x\}|.$$

Hammersley [32] first proved that

$$A_n \approx \mu^n, \tag{6.2}$$

i.e., that the connective constant for SAP's is the same as for SAW's. At first this may seen surprising; however, one can think of A_n/C_n as the probability that a self-avoiding walk is at the origin at time n. In analogy with the case of simple random walk one might then expect

$$\frac{A_n}{C_n} \approx n^{-\delta}, \tag{6.3}$$

for some δ. Note that (6.3) and Proposition 6.2.1 imply (6.2). This is only heuristic, however, and the known proofs of (6.2) are not strong enough to conclude (6.3). This δ is one of a number of "critical exponents" for SAW's about which much is known heuristically and numerically, but for which little is known rigorously. A number of these exponents are discussed in the next section.

6.3 Critical Exponents

Consider C_n, the number of SAW's of length n. By Proposition 6.2.1,

$$C_n = \mu^n r(n),$$

where

$$\lim_{n \to \infty} [r(n)]^{1/n} = 1.$$

We would like to have more precise information about the function $r(n)$. Consider first the case of simple random walk. There seems to be no non-trivial analogue of $r(n)$ for simple random walk since $|\Lambda_n| = (2d)^n$ exactly. However, let

$$\tilde{r}(n) = \frac{r(2n)}{r(n)r(n)} = \frac{C_{2n}}{C_n C_n}.$$

The right hand side can be interpreted as the probability that two independent SAW's of length n can be put together to form a SAW of length $2n$ (where the probability is over the uniform probability measure on SAW's), i.e., the probability that two SAW's of length n have no points in common other than the origin. The analogue of this probability for simple random walk is the function $f(n)$ studied in Chapters 3-5. Recall (Theorems 4.12 and 5.20)

$$f(n) \approx \begin{cases} n^{-\zeta}, & d < 4, \\ (\ln n)^{-1/2}, & d = 4, \\ c, & d > 4. \end{cases}$$

In analogy we would expect

$$\tilde{r}(n) \approx \begin{cases} n^{-\tilde{\zeta}}, & d < 4, \\ (\ln n)^{-a}, & d = 4, \\ c, & d > 4. \end{cases} \tag{6.4}$$

The exponent $\tilde{\zeta}$ is usually denoted $\gamma - 1 = \gamma_d - 1$, and (6.4) suggests

$$r(n) \approx \begin{cases} n^{\gamma-1}, & d < 4, \\ (\ln n)^a, & d = 4, \\ c, & d > 4. \end{cases} \tag{6.5}$$

Recall from Section 5.1 that the conjectured values for ζ are

$$\zeta_2 = \frac{5}{8}, \quad \zeta_3 \simeq .28 \text{ or } .29.$$

Intuitively, one would expect SAW's to be "thinner" than simple random walk paths and hence $\tilde{r}(n) \geq f(n)$. This intuition agrees with the conjectured values for γ [56, 57, 30, 48],

$$\gamma_2 - 1 = \frac{11}{32}, \quad \gamma_3 - 1 \simeq .16.$$

In the critical dimension $d = 4$ it has been conjectured [29] that $a = 1/4$. We should comment that while we have defined the exponent γ by (6.5), there is no proof that such a γ exists. (In contrast, the exponent ζ is known to exist by Theorem 5.3.1.) We will define the other critical exponents for SAW's similarly in this chapter with the understanding that there is no proof that any of the exponents exist.

As mentioned in the previous section, the number of self-avoiding polygons of length n, A_n, is expected to satisfy $A_n/C_n = n^{-\delta}$. We define the exponent $\alpha = \alpha_d$ (which is sometimes referred to as α_{sing}) to be $1 + \gamma - \delta$, i.e.,

$$\mu^{-n} A_n \approx n^{\alpha-2}.$$

A SAP is a SAW with the restriction that $\omega(n) = 0$. If for fixed $x \in Z^d$ we let $C_n(x)$ be the number of SAW's of length n with $\omega(n) = x$, it is similarly expected that

$$\mu^{-n} C_n(x) \approx n^{\alpha-2},$$

assuming, of course, that $n \leftrightarrow x$. We will give the conjectures for α in terms of the exponent ν defined next.

The exponent ν concerns the distribution of the endpoint of the SAW. Let $U = U_n$ denote the uniform probability measure on Γ_n and $\langle\cdot\rangle_U$ expectations with respect to U. Then the mean square displacement exponent ν is defined by

$$\langle|\omega(n)|^2\rangle_U \approx n^{2\nu}.$$

Note that if $P = P_n$ is the uniform measure on Λ_n (simple random walk), then

$$\langle|\omega(n)|^2\rangle_P = n,$$

and hence the exponent ν is equal to $1/2$ for simple walks. Flory [26, 27] gave an argument that predicted for the SAW

$$\nu = \begin{cases} 3/(d+2), & d \leq 4, \\ 1/2, & d > 4. \end{cases}$$

This conjecture suggests that the self-avoidance constraint is not very significant above the critical dimension 4. This is quite plausible since with positive probability the paths of simple random walks do not intersect for $d > 4$. We point out, however, that this plausibility argument is a long way from any kind of proof—even in high dimensions, the number of SAW's is an exponentially small fraction of the number of simple walks. The Flory conjecture gives the correct answer for $d = 1$ (where the SAW problem is particularly easy!), and the $d = 2$ conjecture is expected to be exactly correct. Numerical simulations [54] suggest, however, that the conjecture is not exactly correct in three dimensions, but rather that $\nu_3 = .59...$ In the critical dimension $d = 4$, a logarithmic correction is predicted,

$$n^{-1}\langle|\omega(n)|^2\rangle_U \approx (\ln n)^{1/4}.$$

Slade has proved that the conjecture is correct in high dimensions. (As this book was going to press, Slade and Hara announced that they have proved the following theorem for $d_0 = 5$.)

Theorem 6.3.1 *[62, 63] There exists a $d_0 < \infty$, such that for all $d \geq d_0$, $\nu = 1/2$ and $\gamma = 1$. Moreover, the distribution of $n^{-1/2}\omega(n)$ under U converges to a Gaussian distribution.*

There is a conjectured relationship between ν and α. Consider the number of SAP's of length $2n$, A_{2n}. Then

$$\mu^{-2n}A_{2n} \approx n^{\alpha-2}.$$

A SAP of length $2n$ can be thought of as composed of two SAW's of length n, ω_1 and ω_2, with $\omega_1(n) = \omega_2(n)$ and the restriction that

$$\omega_1(i) \neq \omega_2(j), 0 < i, j < n$$

(here ω_2 is the second half of ω "traversed backwards"). One would expect that the effect of the restriction would be greatest at points near 0 and points near $\omega_1(n) = \omega_2(n)$. Each of these effects should contribute a factor of about $\tilde{r}(n)$. Hence we guess that the number of SAP's ω of length $2n$ with $\omega(n) = x$ is approximately

$$C_n(x)^2\tilde{r}(n)^2,$$

and A_{2n} is approximated by

$$[\sum_{x \in Z^d} C_n(x)^2]\tilde{r}(n)^2.$$

What does a typical term in this sum look like? A typical SAW of length n should be about distance n^ν from the origin. Since $\mu^{-n}C_n \approx n^{\gamma-1}$, and there are on the order of $n^{d\nu}$ points at a distance n^ν from the origin, one would guess for a typical x with $|x| \approx n^\nu$,

$$\mu^{-n}C_n(x) \approx n^{\gamma-1}n^{-d\nu}.$$

Therefore,

$$\mu^{-2n}A_{2n} \approx \sum_{|x|\approx n^\nu} (n^{\gamma-1}n^{-d\nu})^2(n^{1-\gamma})^2 \approx n^{-d\nu}.$$

This gives the conjecture

$$\alpha - 2 = -d\nu,$$

which is sometimes referred to as hyperscaling. If we combine this relation with the conjectures for ν we get

$$\alpha_2 = \frac{1}{2}; \quad \alpha_3 \simeq .23; \quad \alpha_d = 2 - \frac{d}{2}, d \geq 4,$$

with a logarithmic correction in $d = 4$. While the existence of α has not been proved, Madras [52] has proved that if the exponent exists it must satisfy

$$\alpha_2 \leq \frac{5}{2}; \quad \alpha_3 \leq 2; \quad \alpha_d < 2, d \geq 4.$$

We now consider the SAW analogue of the Green's function. Recall for simple random walk,

$$G(0, x) = \sum_{n=0}^{\infty} P\{S_n = x\} = \sum_{n=0}^{\infty} (2d)^{-n} b_n(x),$$

where $b_n(x) = |\{\omega \in \Lambda_n : \omega(n) = x\}|$. For the self-avoiding walk we define

$$\overline{G}(0, x) = \sum_{n=0}^{\infty} \mu^{-n} C_n(x).$$

We define the exponent $\eta = \eta_d$ by

$$\overline{G}(0, x) \approx |x|^{-(d-2+\eta)}.$$

Thus, η measures the amount the exponent for $\overline{G}$ differs from that of G (recall from Theorem 1.5.4 that $G(0, x) \sim a_d |x|^{2-d}$ for $d \geq 3$). There is a heuristic scaling formula which relates η to γ and ν which we now derive. If $|x|$ is much larger than n^ν then $C_n(x)$ should be negligible. If $|x|$ is of order n^ν then $\mu^{-n} C_n(x) \approx n^{\gamma-1} n^{-d\nu}$ (since there are order $n^{d\nu}$ such points). Therefore, we expect

$$\overline{G}(0, x) \quad \approx \quad \sum_{n \approx |x|^\nu} \mu^{-n} C_n(x)$$

$$\approx \quad \sum_{n=|x|^\nu}^{\infty} n^{\gamma-1} n^{-d\nu} \approx |x|^{(\gamma-d\nu)/\nu}.$$

Therefore $-(d - 2 + \eta) = (\gamma - d\nu)/\nu$, or

$$\gamma = \nu(2 - \eta).$$

One final exponent is defined by considering $C_{n,n}$, the number of pairs of n-step SAW's (ω_1, ω_2) with $\omega_1(0) = 0$, $\omega_2(0)$ arbitrary, and such that

$\omega_1 \cap \omega_2 \neq \emptyset$, i.e., for some i, j, $\omega_1(i) = \omega_2(j)$. The exponent $\Delta = \Delta(d)$ (which is sometimes referred to as Δ_4) is defined by

$$\mu^{-2n} C_{n,n} \approx n^{2\Delta + \gamma - 2}.$$

We will relate this exponent to the other exponents. Note that there are C_n choices for ω_1; C_n choices for ω_2 up to translation; and $(n+1)^2$ choices for the pair (i, j) such that $\omega_1(i) = \omega_2(j)$ (once (i, j) is chosen, the starting point $\omega_2(0)$ is determined). Therefore, if we did not have to worry about overcounting, we would have $C_{n,n} = (n+1)^2 C_n^2$. However, a given pair (ω_1, ω_2) may intersect in a number of points; in fact, we would expect the number of intersections to be on the order of b_n, the expected number of intersections of two SAW's starting at the origin. We then get

$$C_{n,n} [C_n]^{-2} \approx n^2 b_n^{-1},$$

or,

$$\mu^{-2n} C_{n,n} \approx n^{2\gamma} b_n^{-1}. \qquad (6.6)$$

We now try to estimate b_n (this is the SAW analogue of (3.4) for simple random walk). Consider two independent SAW's, ω_1 and ω_2 starting at the origin. Since a walk of length n goes distance about n^ν, the number of points visited in the ball of radius m should look like $m^{1/\nu}$. Therefore, the probability that a point x is hit by ω_1 should look like $|x|^{\frac{1}{\nu} - d}$ and the expected number of x which are hit by both ω_1 and ω_2 should look like

$$\sum_{|x| \approx n^\nu} |x|^{2(\frac{1}{\nu} - d)} \approx n^{2 - \nu d}, \quad d = 2, 3.$$

Therefore $b_n \approx n^{2 - \nu d}$ for $d = 2, 3$ and from (6.6),

$$\mu^{-2n} C_{n,n} \approx n^{d\nu + 2\gamma - 2}, \quad d = 2, 3.$$

Therefore, $2\Delta + \gamma - 2 = d\nu + 2\gamma - 2$ or

$$d\nu = 2\Delta - \gamma, \quad d = 2, 3.$$

6.4   Edwards Model

There are a number of other measures on random walks which favor self-avoiding or almost self-avoiding walks. In this section we discuss two such measures. These are measures on the set of simple walks Λ_n. For any $\omega \in \Lambda_n$, define $J(\omega) = J_n(\omega)$ to be the number of self-intersections, i.e.,

$$J(\omega) = \sum_{0 \leq i < j \leq n} I\{\omega(i) = \omega(j)\} = \frac{1}{2} \sum_{i \neq j} \delta(\omega(i) - \omega(j)).$$

The expected value of J with respect to the simple random walk measure can be estimated easily using Theorem 1.2.1:

$$
\begin{aligned}
\langle J \rangle_P &= \sum_{0 \le i < j \le n} P\{\omega(i) = \omega(j)\} \\
&= \sum_{i=0}^{n-1} \sum_{j=i+1}^{n} p(j-i) \\
&\sim \begin{cases} cn^{3/2}, & d=1, \\ cn \ln n, & d=2, \\ cn, & d \ge 3. \end{cases}
\end{aligned}
$$

For any $\beta \ge 0$, we let $U^\beta = U_n^\beta$ be the probability measure on Λ_n given by

$$
U^\beta(\omega) = \frac{\exp\{-\beta J(\omega)\}}{\langle \exp\{-\beta J\} \rangle_P}.
$$

Note that $\beta = 0$ corresponds to the simple random walk and the $\beta \to \infty$ limit gives the self-avoiding walk. This measure is called the *weakly self-avoiding walk* or the *Domb-Joyce model*. It is conjectured that for every $\beta > 0$, this measure is in the same "universality class" as the usual self-avoiding walk. What is meant by physicists by this term has never been stated precisely, but one thing that is definitely implied is that the critical exponents for the weakly self-avoiding walk should be the same as for the self-avoiding walk. For example, if $\nu = \nu(\beta, d)$ is defined by

$$
\langle |\omega(n)|^2 \rangle_{U^\beta} \approx n^{2\nu},
$$

then for any $\beta > 0$, $\nu(\beta, d)$ is expected to the the same as ν_d for the self-avoiding walk.

There is a similar model which is a discrete analogue of a model first defined for continuous time processes. Let B_t be a standard Brownian motion defined on the probability space $(\Omega, \mathcal{F}, P)$, $0 \le t \le 1$. Consider the (formal) self-intersection random variable

$$
V = \int_0^1 \int_0^1 \delta(B_s - B_t)\, ds\, dt,
$$

and for each $\beta \ge 0$ define a measure Q_β by

$$
\frac{dQ_\beta}{dP} = \frac{\exp\{-\beta V\}}{\langle \exp\{-\beta V\} \rangle_P}.
$$

Then the set of paths B_t under the measure Q_β is called the *Edwards model* [20]. This is only formal, of course. However, one can approximate the delta function by approximate delta functions δ_ϵ, and then V_ϵ and $Q_{\beta,\epsilon}$ are well defined. Then one would hope to be able to take a weak limit of the measures $Q_{\beta,\epsilon}$. This has been done for $d = 2, 3$ [69, 72].

What is the random walk analogue of this model? Consider a simple random walk of length n, ω. Then $\tilde{B}_t = n^{-1/2}\omega([nt])$ is approximately a Brownian motion. Since the steps are of size $n^{-1/2}$, we approximate the delta function by

$$\delta^n(x) = \delta^n(x^1, \ldots, x^d) = \begin{cases} n^{d/2}, & |x^i| \leq \frac{1}{2}n^{-1/2}, \\ 0, & \text{otherwise.} \end{cases}$$

We then approximate V by

$$\int_0^1 \int_0^1 \delta(B_s - B_t)\,ds\,dt \;\simeq\; (\frac{1}{n})^2 \sum_{i=0}^n \sum_{j=0}^n \delta^n(\tilde{B}_{i/n} - \tilde{B}_{j/n})$$

$$= \; n^{(d-4)/2} \sum_{i=0}^n \sum_{j=0}^n \delta(\omega(i) - \omega(j))$$

$$= \; n^{(d-4)/2}(2J(\omega) + (n+1)).$$

Let

$$\bar{J} = 2n^{(d-4)/2}(J - \langle J \rangle_P).$$

Then the *(discrete) Edwards model* is the measure $Q^\beta = Q_n^\beta$ given by

$$Q^\beta(\omega) = \frac{\exp\{-\beta\bar{J}\}}{\langle \exp\{-\beta\bar{J}\}\rangle_P}.$$

Note that adding a constant to the random variable $\bar{J}$ does not change Q^β, so we may use $\bar{J}$ for convenience. If $d = 4$, the Edwards model is the same as the weakly self-avoiding walk (more precisely, $Q^\beta = U^{2\beta}$) while for $d = 2, 3$ the Edwards model interaction is significantly weaker than that in the weakly self-avoiding walk.

We will consider the Edwards model in two dimensions. For $d = 2$,

$$\bar{J} = \frac{2}{n}(J - \langle J \rangle_P).$$

Since $\langle J \rangle_P \sim cn \ln n$, one might expect that $\bar{J}$ gets large with n. However, the major contribution to J is from intersections with $|i - j|$ small, and the number of such short-range intersections is relatively uniform from path to path. The contribution from long-range intersections turns out to be of order n, and it is this contribution which is important in Q^β.

Proposition 6.4.1 *If $d = 2$,*

$$\mathrm{Var}(J) = \langle J^2 \rangle_P - \langle J \rangle_P^2 = O(n^2),$$

and hence

$$\mathrm{Var}(\overline{J}) \leq c.$$

Proof. If we expand the sums in the definition of the variance we get

$$\mathrm{Var}(J) \leq 2 \sum_{(i_1, j_1, i_2, j_2) \in A} q(i_1, j_1, i_2, j_2),$$

where $q(i_1, j_1, i_2, j_2)$ equals

$$P\{\omega(i_1) = \omega(j_1), \omega(i_2) = \omega(j_2)\} - p(j_1 - i_1)p(j_2 - i_2),$$

and

$$A = \{0 \leq i_1, i_2, j_1, j_2 \leq n; i_1 < j_1; i_2 < j_2; i_1 \leq i_2\}.$$

We partition A into three sets:

$$A^1 = A \cap \{i_1 < j_1 \leq i_2 < j_2\},$$
$$A^2 = A \cap \{i_1 \leq i_2 < j_2 \leq j_1\},$$
$$A^3 = A \cap \{i_1 \leq i_2 \leq j_1 < j_2\}.$$

If $i_1 < j_1 \leq i_2 < j_2$, then $q(i_1, j_1, i_2, j_2) = 0$, so the sum over A^1 is zero. If $i_1 \leq i_2 < j_2 \leq j_1$,

$$P\{\omega(i_1) = \omega(j_1), \omega(i_2) = \omega(j_2)\} = p(j_2 - i_2)p((j_1 - i_1) - (j_2 - i_2)).$$

Therefore, using Theorem 1.2.1,

$$
\begin{aligned}
q(i_1, j_1, i_2, j_2) &= p(j_2 - i_2)[p((j_1 - i_1) - (j_2 - i_2)) - p(j_1 - i_1)] \\
&\leq c(j_1 - i_1)^{-1}((j_1 - i_1) - (j_2 - i_2) + 1)^{-1}.
\end{aligned}
$$

If we set $k_1 = i_2 - i_1, k_2 = j_2 - i_2, k_3 = j_1 - j_2$, then the sum over A^2 is bounded by a constant times

$$(n+1) \sum_{k_1=0}^{n} \sum_{k_2=1}^{n} \sum_{k_3=0}^{n} (k_1 + k_2 + k_3)^{-1}(k_1 + k_3 + 1)^{-1}$$

$$\leq cn \sum_{k_1=0}^{n} \sum_{k_3=0}^{n} \ln(\frac{k_1 + k_3 + n}{k_1 + k_3 + 1})(k_1 + k_3 + 1)^{-1}$$

$$\leq cn \sum_{j=0}^{2n} (j+1) \ln(\frac{j+1+n}{j+1})(j+1)^{-1}$$

$$= O(n^2).$$

Finally, if $i_1 \leq i_2 \leq j_1 < j_2$,

$$P\{\omega(i_1) = \omega(j_1), \omega(i_2) = \omega(j_2)\}$$
$$= P\{\omega(i_1) = \omega(j_1)\}P\{\omega(i_2) = \omega(j_2) \mid \omega(i_1) = \omega(j_1)\}$$
$$\leq c(j_1 - i_1 + 1)^{-1}(j_2 - j_1)^{-1},$$

and hence

$$q(i_1, j_1, i_2, j_2) \leq c(j_1 - i_1 + 1)^{-1}[(j_2 - j_1)^{-1} - (j_2 - i_2)^{-1}].$$

Similarly,

$$q(i_1, j_1, i_2, j_2) \leq c(j_2 - i_2)^{-1}[(i_2 - i_1 + 1)^{-1} - (j_1 - i_1 + 1)^{-1}].$$

Therefore, if $k_1 = i_2 - i_1, k_2 = j_1 - i_2, k_3 = j_2 - j_1$, the sum over A^3 is bounded by a constant times

$$(n+1)\{\sum_{k_1=0}^{n} \sum_{k_2=0}^{n} \sum_{k_3=k_1+1}^{n} (k_1 + k_2 + 1)^{-1}[k_3^{-1} - (k_3 + k_2)^{-1}]$$

$$+ \sum_{k_1=0}^{n} \sum_{k_2=0}^{n} \sum_{k_3=1}^{k_1} (k_2 + k_3)^{-1}[(k_1 + 1)^{-1} - (k_1 + k_2 + k_3)^{-1}]\}.$$

If we sum over k_3 in the first triple sum, we see that this triple sum is bounded by a constant times

$$\sum_{k_1=0}^{n} \sum_{k_2=0}^{n} (k_1 + k_2 + 1)^{-1}(\ln(k_1 + k_2 + 1) - \ln(k_1 + 1))$$

$$\leq \sum_{k_1=0}^{n} \sum_{j=k_1+1}^{2n+1} j^{-1}(\ln j - \ln(k_1 + 1))$$

$$= \sum_{j=1}^{2n+1} j^{-1} \sum_{k_1=0}^{j-1} (\ln j - \ln(k_1 + 1))$$

$$= O(n).$$

Similarly, the second triple sum is $O(n)$, and hence the sum over A^3 is $O(n^2)$. This proves the proposition. □

The proposition shows that $\mathrm{Var}(\bar{J})$ is bounded for $d = 2$. With sharper estimates, see e.g. Stoll [68], one can show that for every $\beta > 0$,

$$\langle \exp\{-\beta \bar{J}\}\rangle_P \leq c(\beta) < \infty. \tag{6.7}$$

This implies that the measure Q^β is "absolutely continuous" with respect to P (in fact, Varadhan [69] proved that the two dimensional continuous Edwards model is absolutely continuous with respect to Wiener measure). One consequence of (6.7) is that the discrete Edwards model is not in the same universality class as the self-avoiding walk or weakly self-avoiding walk in two dimensions. To see this, consider the mean-square displacement exponent ν,

$$\langle |\omega(n)|^2 \rangle_{Q^\beta} \approx n^{2\nu}.$$

By Hölder's inequality and (6.7),

$$
\begin{aligned}
\langle |\omega(n)|^2 \exp\{-\beta \overline{J}\} \rangle_P &\leq \langle |\omega(n)|^4 \rangle_P^{1/2} \langle \exp\{-2\beta \overline{J}\} \rangle_P^{1/2} \\
&\leq c_\beta n.
\end{aligned}
$$

Therefore, using Proposition 6.4.1 and (6.7),

$$c_1(\beta) n \leq \langle |\omega(n)|^2 \rangle_{Q^\beta} \leq c_2(\beta) n,$$

and hence $\nu = 1/2$, which is not the conjectured value for the self-avoiding walk.

If $d = 3$, a similar argument shows that $\mathrm{Var}(J) \asymp n^{1/2}$ and hence that $\overline{J}$ does get large as n grows. Therefore, the measure Q^β becomes "singular" with respect to the measure P as $n \to \infty$. Westwater [72] has proved that the three dimensional (continuous) Edwards model is singular with respect to Wiener measure. In two dimensions it is known that the continuum limit of the discrete Edwards model is the (continuous) Edwards model [68]; it is certainly expected that this is true in three dimensions, but it has not been proved.

6.5 Kinetically Growing Walks

The self-avoiding walk, weakly self-avoiding walk, and the Edwards model are examples of "configurational" measures on random walks paths. Such measures are natural from the viewpoint of equilibrium statistical mechanics. In these measures walks which minimize "energy" are favored, where the energy is some function of the number of self-intersections.

These configurational measures are not natural if one wants to consider a random walk as a stochastic process. In particular, these measures on Λ_n or Γ_n are not consistent. We say that a sequence of measures λ_n on Λ_n is consistent if for every $\omega \in \Lambda_n$, $m \geq 0$,

$$\lambda_n(\omega) = \sum_{\omega \prec \eta} \lambda_{n+m}(\eta),$$

where $\omega \prec \eta$ means that η extends ω, i.e., $\eta(i) = \omega(i), 0 \le i \le n$. Whenever a sequence of consistent measures λ_n on Λ_n is given, there is a well-defined measure λ on the space of infinite random walk paths defined on cylinder sets by λ_n. It is easy to see that the self-avoiding measures U_n are not consistent; in fact, one can find SAW's ω that are "trapped", i.e., such that ω cannot be extended to any longer SAW.

One way to describe a consistent set of measures is to give transition probabilities. Suppose λ_n are consistent probability measures on Λ_n. Then if $\omega \in \Lambda_n, \tilde{\omega} \in \Lambda_{n+1}, \omega \prec \tilde{\omega}$, let

$$\pi(\tilde{\omega} \mid \omega) = \frac{\lambda_{n+1}(\tilde{\omega})}{\lambda_n(\omega)}.$$

Then,

$$\lambda_n(\omega) = \prod_{i=0}^{n-1} \pi(\omega^{i+1} \mid \omega^i), \tag{6.8}$$

where ω^i denotes the unique walk in Λ_i with $\omega^i \prec \omega$. Conversely, if the transition probabilities $\pi(\tilde{\omega} \mid \omega)$ are given, we can define consistent probability measures λ_n by (6.8). The transitions π can be viewed as Markovian transition probabilities on the state space $\Lambda = \cup_{n=0}^\infty \Lambda_n$ or as non-Markovian transitions on the state space Z^d. Such walks are often referred to as *kinetically growing walks*.

The first attempt to define a "kinetically growing self-avoiding walk" might be to let a random walker choose randomly among all sites that it has not visited. Let $V_n(x)$ be the number of visits to x,

$$V_n(x) = V_n(x, \omega) = |\{j : \omega(j) = x\}|.$$

Then such a walk would correspond to transitions

$$\pi\{\tilde{\omega}(n+1) = x | \omega\} = \frac{1 - V_n(x)}{\sum_{|y-\omega(n)|=1}(1 - V_n(y))}, \quad |x - \omega(n)| = 1$$

(note that since $\pi(\tilde{\omega}|\omega)$ is non-zero only for $\omega \prec \tilde{\omega}$, it suffices in defining transitions to give the probability that $\tilde{\omega}(n + 1) = x$). Unfortunately, it is not difficult to show that a walker using these transitions will eventually get "trapped" so that $V_n(x, \omega) = 1$ for each $|x - \omega(n)| = 1$. We can do a "weak" version of this idea, however, by discouraging self-intersections but not forbidding them. Let $\beta > 0$ and choose transitions

$$\pi\{\tilde{\omega}(n+1) = x \mid \omega\} = \frac{\exp\{-\beta V_n(x)\}}{\sum_{|y-\omega(n)|=1} \exp\{-\beta V_n(x)\}}, \quad |x - \omega(n)| = 1.$$

We call this the *myopic self-avoiding walk*. Here "myopic" emphasizes the fact that the walker only looks at its nearest neighbors when choosing the next step. This has also been labelled as the "true" self-avoiding walk [3]. Does this process look qualitatively like the usual self-avoiding walk? While no rigorous work has been done, heuristic arguments and numerical work suggest that it is significantly different then the usual self-avoiding walk. For example, the mean-square displacement of the myopic self-avoiding walk is expected to grow like $n^{4/3}$ in one dimension and n for $d \geq 2$ (with possible logarithmic corrections in two dimensions). This shows that two is the critical dimension for this process rather than four.

This example shows that it may be difficult to find a kinetically growing walk which is qualitatively like the usual self-avoiding walk. There is, however, a natural way to try to define such a walk. For any SAW $\omega \in \Gamma_n$, let

$$C_{n+m}(\omega) = |\{\eta \in \Gamma_{n+m} : \omega \prec \eta\}|,$$

and

$$\tilde{U}_n(\omega) = \lim_{m \to \infty} \frac{C_{n+m}(\omega)}{C_{n+m}}, \qquad (6.9)$$

assuming the limit exists. It is easy to verify that the $\tilde{U}_n$ are consistent, with transitions

$$\pi(\tilde{\omega} \mid \omega) = \frac{\tilde{U}_{n+1}(\tilde{\omega})}{\tilde{U}_n(\omega)} = \lim_{m \to \infty} \frac{C_m(\tilde{\omega})}{C_m(\omega)}. \qquad (6.10)$$

This walk is called the *infinite self-avoiding walk*. The problem is that the limit in (6.9) has not been proven to exist, except in high dimensions [44], although it certainly should exist. It can be shown [51] that

$$\liminf_{m \to \infty} \frac{C_{n+m}(\omega)}{C_{n+m}} > 0,$$

for any $\omega \in \Gamma_n$ that is not "trapped".

We now replace (6.10) with a limit that is known to exist. Intuitively, $\pi(\tilde{\omega} \mid \omega)$ is the probability that a long self-avoiding walk starting at $\omega(n)$, conditioned to avoid ω, has initial step $\tilde{\omega}(n+1)$. Let us replace "self-avoiding walk" with "simple random walk" in the last sentence. For ease, assume $d \geq 3$. Then we can define the transitions

$$\pi\{\tilde{\omega}(n+1) = x \mid \omega\} = \frac{\mathrm{Es}_\omega(x)}{\sum_{|y-\omega(x)|=1} \mathrm{Es}_\omega(y)}, \quad |x - \omega(n)| = 1, x \notin \omega.$$

Here we write ω for the set $\{\omega(0), \ldots, \omega(n)\}$ and $\mathrm{Es}_\omega(\cdot)$ is as defined in Section 2.2. Since $\mathrm{Es}_\omega(\cdot)$ is the harmonic function on ω^c with boundary value 0 on ω and 1 at infinity, this walk has been labelled the *Laplacian random walk*. It turns out that this walk is equivalent to the random walk obtained by "erasing loops" from simple random walk, hence the walk is also called the *loop-erased walk*. This walk can also be defined in two dimensions by taking appropriate limits.

A number of rigorous results are known about the Laplacian random walk and these will be derived in the next chapter. Since we can get some rigorous results about this process, it would be nice to know if the Laplacian walk is in the same universality class as the infinite self-avoiding walk defined by (6.10), which we believe is in the same universality class as the usual self-avoiding walk. Intuitively, a self-avoiding walk should be "thinner" than a simple random walk and hence more likely to avoid sets. Therefore, one might guess that the Laplacian walk weights more heavily than the infinite self-avoiding walk those directions which move "away from ω". If anything, this should force the paths of the Laplacian walk to go to infinity faster. In fact, the results of the next section and numerical work show that this is the case. Consider the mean square displacement exponent, ν for the Laplacian random walk. Then it is conjectured [31, 42] that: $\nu_2 = 4/5$; $\nu_3 \simeq .62$; $\nu_d = 1/2$ for $d \geq 4$ with logarithmic correction $(\ln n)^{1/3}$ in four dimensions. In the next chapter we will derive the rigorous results: $\nu_2 \geq 3/4$; $\nu_3 \geq 3/5$; $\nu_d = 1/2$ for $d \geq 4$ with the logarithmic correction in four dimensions being between $(\ln n)^{1/3}$ and $(\ln n)^{1/2}$. The Laplacian random walk is therefore in a different universality class from the usual self-avoiding walk, at least if the conjectures about the usual self-avoiding walk are true.

A one parameter family of Laplacian random walks, indexed by $s > 0$, can be defined by the transitions [50]

$$\pi\{\tilde{\omega}(n+1) = x \mid \omega\} = \frac{[\mathrm{E}s_\omega(x)]^s}{\sum_{|y-\omega(x)|=1}[\mathrm{E}s_\omega(y)]^s}, \quad |x - \omega(n)| = 1.$$

As $s \to \infty$, the paths become straight lines. The case $s \to 0$ has been labelled the indefinitely growing self-avoiding walk (in this case, the walker chooses randomly among all nearest neighbors that will not eventually "trap" the walker). It is expected that the mean square displacement exponent varies continuously with s.

6.6 Monte Carlo Simulations

Because the estimation of critical exponents for self-avoiding walks has proven to be too hard a problem to answer rigorously, computer simulations are useful in predicting behavior. This leads to an interesting problem in itself: how does one generate SAW's on a computer? Since the number of SAW's of length n grows exponentially in n, we cannot expect to write down all SAW's of a given length except for small values of n. We therefore would like to do Monte Carlo simulations, i.e., sampling from the uniform distribution on Γ_n. Sampling from simple random walk is easy and it takes only order n operations to produce a walk of length n. It would not be practical, of course, to generate SAW's by generating simple walks and discarding those that are not self-avoiding since the average number of operations needed to produce a SAW of length n would grow exponentially in n. Trying to modify such an algorithm by choosing steps so that a walk tries to stay self-avoiding can lead to probability distributions significantly different than that of the uniform measure on self-avoiding walks (see Section 6.5).

The most efficient algorithms for generating walks come from performing a Markov chain on the set of SAW's. Suppose Π is a transition matrix for a discrete time Markov chain X_j on the countable state space Ω. Suppose that Π is ergodic (for every $x, y \in \Omega$, $\Pi^j(x, y) > 0$ for some j) and aperiodic (which will be guaranteed for ergodic Π if $\Pi(x, x) > 0$ for some $x \in \Omega$). Let λ be a stationary probability measure for Π, i.e., for each $x \in \Omega$,

$$\sum_{y \in \Omega} \lambda(y)\Pi(y, x) = \lambda(x). \tag{6.11}$$

Then it is well known that for any $x \in \Omega$, the distribution of $\{X_j \mid X_0 = x\}$ approaches λ as $j \to \infty$. This suggests that if we choose a large N and start with $X_0 = x$, then

$$X_N, X_{2N}, X_{3N}, \ldots$$

will be approximately independent samples from the distribution λ. Of course, how large N must be depends on Π. We will consider two examples where Π is symmetric with respect to λ,

$$\Pi(x, y)\lambda(x) = \Pi(y, x)\lambda(y) \tag{6.12}$$

(this is sometimes called the detailed balance condition). If Π satisfies (6.12) for a given λ, then λ is an invariant measure for Π. The rate of convergence to equilibrium will be controlled by the second largest eigenvalue for the operator Π, a_1, which is given by the variational formula (see e.g. [46])

$$1 - a_1 = \inf \frac{\sum_{x, y \in \Omega} h(x)\Pi(x, y)\lambda(x)h(y)}{\sum_{x \in \Omega} h(x)^2 \lambda(x)},$$

where the infimum is over all h with $\sum h(x)\lambda(x) = 0$ and $\sum h(x)^2 \lambda(x) < \infty$. If Ω is finite, a_1 is always positive for ergodic Π while a_1 may or may not be positive for infinite Ω.

In the first example [5] the state space Ω will be the set of all finite SAW's, $\Gamma = \cup_{n=0}^{\infty} \Gamma_n$. Let $|\omega|$ denote the length of ω. Let $\beta > 0$ and let Π be given by

$$\Pi(\omega, \tilde{\omega}) = \begin{cases} \beta(1 + 2d\beta)^{-1}, & |\tilde{\omega}| = |\omega| + 1, \omega \prec \tilde{\omega}, \\ (1 + 2d\beta)^{-1}, & |\omega| = |\tilde{\omega}| + 1, \tilde{\omega} \prec \omega, \\ r(\omega), & \omega = \tilde{\omega} \\ 0, & \text{otherwise,} \end{cases}$$

where $r(\omega)$ is chosen so that

$$\sum_{\tilde{\omega} \in \Omega} \Pi(\omega, \tilde{\omega}) = 1.$$

Consider the measure $\overline{\lambda}(\omega) = \beta^{|\omega|}$. Then Π satisfies (6.12) for $\overline{\lambda}$. Note if $\beta < \mu^{-1}$, then

$$Z(\beta) \doteq \sum_{\omega \in \Omega} \overline{\lambda}(\omega) = \sum_{n=0}^{\infty} C_n \beta^n < \infty,$$

and hence we can define an invariant probability measure by

$$\lambda(\omega) = [Z(\beta)]^{-1} \overline{\lambda}(\omega).$$

The measure $\lambda = \lambda_\beta$ is sometimes called the grand canonical ensemble for SAW's. Note that if $\beta = \beta_n = \mu^{-1} - (1/n)$, then

$$\langle |\omega| \rangle_{\lambda_\beta} \approx n$$

(at least if $\mu^{-n} C_n \approx n^{\gamma-1}$ as expected). It is therefore reasonable to assume that the qualitative behavior of the grand canonical ensemble at β_n should be the same as the SAW of length n (see Theorems 2.4.2 and 2.4.3). It is impossible to prove rigorously what the rate of convergence for this algorithm is without detailed knowledge of SAW's (which is what the algorithm is trying to discover). However, it has been proved that if $\mu^{-1} C_n \approx n^{\gamma-1}$, then N of order $n^{1+\gamma}$ is large enough to get approximately independent samples [64].

Another choice for Π which has proved to be extremely efficient is the pivot algorithm [54]. In this case the state space is $\Omega = \Gamma_n$ for some n. Let $\mathcal{O}$ be the set of d-dimensional orthogonal transformations which leave Z^d invariant. For $d = 2$, this consists of rotations by integral multiples of $\pi/2$, reflections about the coordinate axes, and reflections about the diagonals. In all dimensions, $\mathcal{O}$ is finite. If $T \in \mathcal{O}$ and $\omega \in \Gamma_k$, then $T\omega \in \Gamma_k$. The pivot algorithm goes as follows. Start with $\omega \in \Gamma_n$. Choose a random k uniformly on $\{0, \ldots, n-1\}$ and a $T \in \mathcal{O}$, again uniformly. Consider the walk $\tilde{\omega}$ obtained by fixing the first k steps of ω and then performing the transformation T on the remaining $n - k$ steps (considering $\omega(k)$ as the origin). This new walk may or may not be self-avoiding. If it is we choose this $\tilde{\omega}$; otherwise we stay with the SAW ω. This algorithm corresponds to the matrix Π with $\Pi(\omega, \tilde{\omega}) = (n|\mathcal{O}|)^{-1}$ if $\tilde{\omega}$ can be obtained from ω by such a pivot transformation and $\Pi(\omega, \omega)$ chosen so that Π is a stochastic matrix. Since $\Pi(\omega, \tilde{\omega}) = \Pi(\tilde{\omega}, \omega)$, Π satisfies (6.2) with λ the uniform probability on Γ_n. It can be shown that Π is ergodic and aperiodic.

At first this algorithm would seem inefficient since only a small fraction of the pivot transformations are allowed. However, it is expected that the probability that a transformation produces a SAW is of order $n^{-\rho}$ where ρ is not very large. Also, when a pivot transformation is accepted, a significant change is made in the SAW (as opposed to the previous algorithm where it takes a large number of moves to make significant changes). Again, it is impossible to give rigorous bounds on the rate of convergence. However, for practical purposes in calculating certain exponents, e.g., the mean square exponent ν, "effectively independent" samples from Γ_n seem to be obtained from performing the algorithm order n or $n \ln n$ steps.

Similar algorithms have been developed for generating self-avoiding polygons [15, 53].

Chapter 7

Loop-Erased Walk

7.1 Introduction

In this chapter we discuss the loop-erased or Laplacian self-avoiding random
walk. We will primarily use the loop-erased characterization of the walk
because it is the one that allows for rigorous analysis of the model. In
Proposition 7.3.1 we show that this is the same as the Laplacian random
walk defined in Section 6.5.

The loop-erased walk, like the usual self-avoiding walk, has a critical dimension of four. If $d > 4$, then the number of points remaining after erasing
loops is a positive fraction of the total number of points. We prove a strong
law for this fraction and show that the loop-erased process approaches a
Brownian motion. If $d \leq 4$, the proportion of points remaining after erasing
loops goes to zero. In the critical dimension $d = 4$, however, we can still
prove a weak law for the number of points remaining. From this we can
show that the process approaches a Brownian motion for $d = 4$, although a
logarithmic correction to scaling is needed. For $d < 4$, the number of points
erased is not uniform from path to path, and we do not expect a Gaussian
limit.

Sections 7.2-7.4 give the basic properties of the loop-erased walk. The
definition is a little more complicated in two dimensions because simple random walk is recurrent. In Section 7.5 we give upper bounds on the number
of points remaining for $d \leq 4$. This allows us to give lower bounds on
the mean square displacement in two and three dimensions, Theorem 7.6.2.
The final section discusses the walk for $d \geq 4$ and proves the convergence
to Brownian motion.

7.2 Erasing Loops

In this section we describe a procedure which assigns to each finite simple random walk path λ a self-avoiding walk $L\lambda$. Let $\lambda = [\lambda(0), \ldots, \lambda(m)]$ be a simple random walk path of length m. If λ is self-avoiding, let $L\lambda = \lambda$. Otherwise, let

$$
\begin{aligned}
t &= \inf\{j : \lambda(i) = \lambda(j) \text{ for some } 0 \le i < j\}, \\
s &= \text{the } i < t \text{ with } \lambda(i) = \lambda(t),
\end{aligned}
$$

and let $\tilde{\lambda}$ be the $m - (t - s)$ step path

$$
\tilde{\lambda}(j) = \left\{
\begin{array}{ll}
\lambda(j), & 0 \le j \le s, \\
\lambda(j + t - s), & s \le j \le m - (t - s).
\end{array}
\right.
$$

If $\tilde{\lambda}$ is self-avoiding we let $L\lambda = \tilde{\lambda}$. Otherwise, we perform this procedure on $\tilde{\lambda}$ and continue until we eventually obtain a self-avoiding walk $L\lambda$ of length $n \le m$. This walk clearly satisfies $(L\lambda)(0) = \lambda(0)$ and $(L\lambda)(n) = \lambda(m)$.

There is another way to define $L\lambda$ which can easily be seen to be equivalent. Let

$$
s_0 = \sup\{j : \lambda(j) = \lambda(0)\},
$$

and for $i > 0$,

$$
s_i = \sup\{j : \lambda(j) = \lambda(s_{i-1} + 1)\}.
$$

Let

$$
n = \inf\{i : s_i = m\}.
$$

Then

$$
L\lambda = [\lambda(s_0), \lambda(s_1), \ldots, \lambda(s_n)].
$$

The loop-erasing procedure depends on the order of the points. Suppose we wish to erase loops in the reverse direction. More precisely, let

$$
\lambda_R(j) = \lambda(m - j), \quad 0 \le j \le m,
$$

and define reverse loop-erasing L^R by

$$
L^R\lambda = (L\lambda_R)_R.
$$

It is not difficult to construct λ such that $L\lambda \ne L^R\lambda$. However, we prove here that if λ is chosen using the distribution of simple random walk, then $L\lambda$ and $L^R\lambda$ give the same distribution. Recall that Λ_m is the set of simple random walk paths of length m starting at the origin.

Lemma 7.2.1 *For each $m \geq 0$, there exists a bijection $T^m : \Lambda_m \to \Lambda_m$ such that for each $\lambda \in \Lambda_m$,*

$$L(\lambda) = L^R(T^m \lambda).$$

Moreover, λ and $T^m \lambda$ visit the same points with the same multiplicities.

Proof. We will prove by induction on m. The lemma is clearly true for $m = 0, 1$, so assume that such bijections T^k exist for all $k < m$. Let $\lambda \in \Lambda_m$ and assume

$$L\lambda = \gamma = [\gamma(0), \ldots, \gamma(n)].$$

If $n = 0$, we set $T^m \lambda = \lambda$. Assume $n > 0$. Let

$$s_n = \sup\{j : \lambda(j) = 0\},$$

and define $s_0 \leq s_1 \leq \cdots \leq s_{n-1} \leq s_n$ by stating that s_i is the largest integer with $\lambda(s_i) = 0$ and $\{\lambda(0), \ldots, \lambda(s_i)\} \cap \{\gamma(i+1), \ldots, \gamma(n)\} = \emptyset$. We define "loops" K_i, $i = 1, \ldots, n$, as follows: if $s_i = s_{i-1}$ then $K_i = \emptyset$; otherwise, let j be the smallest integer greater than s_{i-1} with $\lambda(j) = \gamma(i)$ and let

$$K_i = [\lambda(j), \lambda(j+1), \ldots, \lambda(s_i - 1), \lambda(s_{i-1}), \lambda(s_{i-1} + 1), \ldots, \lambda(j-1)].$$

Note that K_i starts at $\gamma(i)$, ends at a nearest neighbor of $\gamma(i)$, $K_i \cap \{\gamma(i+1), \ldots, \gamma(n)\} = \emptyset$.

By the second description of the loop-erasing procedure it is easy to see that

$$L[\lambda(s_n + 1), \ldots, \lambda(m)] = [\gamma(1), \ldots, \gamma(n)].$$

Let $\eta = T^{m-(s_n+1)}[\lambda(s_n + 1), \ldots, \lambda(m)]$ (here we have naturally extended T^k to walks which do not start at the origin). Then $\eta = [\eta(0), \ldots, \eta(m - s_n - 1)]$ traverses the same points as $[\lambda(s_n + 1), \ldots, \lambda(m)]$ with the same multiplicities, and $L^R \eta = [\gamma(1), \ldots, \gamma(m)]$. Let

$$t_i = \inf\{j : \eta(j) = \gamma(i)\}.$$

Then we set

$$T^m \lambda = [\lambda(0), \ldots, \lambda(s_0), K_1, \eta(t_1), \ldots, \eta(t_2 - 1), K_2, \eta(t_2), \ldots,$$

$$\eta(t_3 - 1), \ldots, K_n, \eta(t_n), \ldots, \eta(m - s_n - 1)].$$

Clearly $T^m \lambda$ traverses the same points as λ and it is easy to check that $L^R(T^m \lambda) = \gamma$.

To prove that T^m is a bijection we will describe the inverse map. Let $\omega = [\omega(0), \ldots, \omega(m)] \in \Lambda_m$ with $L^R \omega = \gamma$. Let

$$u_i = \inf\{j : \omega(j) = \gamma(i)\}, \quad i = 1, \ldots, n.$$

and $u_{i+1} = m + 1$. For $i = 1, \ldots, n$, if $\omega(j) \neq 0$ for $u_i \leq j \leq u_{i+1}$, then set $v_i = u_i$. Otherwise, let j be the smallest integer less than u_{i+1} with $\omega(j) = 0$ and k the smallest integer greater than j with $\omega(k) = \gamma(i)$ (note such a k must exist since $\omega(u_{i+1} - 1) = \gamma(i)$). We then write

$$\omega = [\omega(0), \ldots, \omega(u_1 - 1), K_1, \omega(v_1), \ldots, \omega(v_2 - 1), K_2, \ldots,$$

$$K_n, \omega(v_n), \ldots, \omega(u_{n+1} - 1)],$$

where

$$K_i = \begin{cases} \emptyset, & \text{if } u_i = v_i, \\ \omega(u_i), \ldots, \omega(v_i - 1), & \text{if } u_i < v_i. \end{cases}$$

With this decomposition it is easy to write the inverse explicitly and verify that it is the inverse. $\square$

7.3 Loop-erased Walk

We will define the loop-erased walk for $d \geq 3$ by erasing loops from the path of an infinite simple random walk. This will be well defined since the random walk is transient. If S_j is a simple random walk in Z^d, $d \geq 3$, let

$$\begin{aligned} u_1 &= \inf\{j : S_j = S_k \text{ for some } 0 \leq k < j\}, \\ v_1 &= \text{the } k < u_1 \text{ with } S_k = S(u_1). \end{aligned}$$

Then set

$$S_j^1 = \begin{cases} S_j, & 0 \leq j \leq v_1, \\ S(j + u_1 - v_1), & v_1 \leq j < \infty \end{cases}$$

We continue inductively by letting

$$\begin{aligned} u_i &= \inf\{j : S_j^{i-1} = S_k^{i-1} \text{ for some } 0 \leq k < j\}, \\ v_i &= \text{the } k < u_i \text{ with } S_k^{i-1} = S^{i-1}(u_i), \end{aligned}$$

and

$$S_j^i = \begin{cases} S_j^{i-1}, & 0 \leq j \leq v_i, \\ S^{i-1}(j + u_i - v_i), & v_i \leq j < \infty. \end{cases}$$

Each S^i has self-intersections; however, we can define

$$\hat{S}(j) = \lim_{i \to \infty} S_j^i,$$

and obtain a self-avoiding walk. We call $\hat{S}$ the *loop-erased self-avoiding random walk* or simply the *loop-erased walk*.

As in the previous section we can use an alternative construction of $\hat{S}$ from S. Let

$$\sigma_0 = \sup\{j : S_j = 0\},$$

and for $i > 0$,

$$\sigma_i = \sup\{j > \sigma_{i-1} : S_j = S(\sigma_{i-1} + 1)\}.$$

Then let

$$\hat{S}(i) = S(\sigma_i).$$

It is easy to check that this is an equivalent definition.

We define the probability measures $\hat{P} = \hat{P}_k$ on Γ_k, the set of SAW's of length k, by

$$\hat{P}(\gamma) = P\{[\hat{S}(0), \ldots, \hat{S}(k)] = \gamma\}.$$

The $\hat{P}_k$ give a consistent set of measures on Γ_k. In the next proposition we show that these measures are the same as for the *Laplacian random walk* which was defined in section 6.5. Therefore, the loop-erased walk and the Laplacian random walk are the same.

Proposition 7.3.1 *If* $\gamma_k = [\gamma(0), \ldots, \gamma(k)] \in \Gamma_k, k \geq 1$, *and* $\gamma_{k-1} = [\gamma(0), \ldots, \gamma(k-1)]$, *then*

$$\hat{P}(\gamma_k) = \hat{P}(\gamma_{k-1})P^{\gamma(k-1)}\{S_1 = \gamma(k) \mid \tau_A = \infty\}$$

$$= \hat{P}(\gamma_{k-1})\frac{\mathrm{Es}_A(\gamma(k))}{\sum_{y \notin A, |y-\gamma(k-1)|=1} \mathrm{Es}_A(y)},$$

where $A = \{\gamma(0), \ldots, \gamma(k-1)\}$ *and as before*

$$\tau_A = \inf\{j \geq 1 : S_j \in A\}.$$

Proof. Let V_m be the event

$$\{\sigma_{k-1} = m, [\hat{S}(0), \ldots, \hat{S}(k-1)] = \gamma_{k-1}\}.$$

Then V_m is the set of paths satisfying

(i) $L[S_0, \ldots, S_m] = \gamma_{k-1}$,

(ii) $S_j \notin A, \ j = m+1, m+2, \ldots$.

Note that (i) and (ii) are conditionally independent given $S_m = \gamma(k-1)$. Also, if $\sigma_{k-1} = m$, then $\hat{S}(k) = S_{m+1}$. Therefore,

$$P\{\hat{S}(k) = \gamma(k) \mid V_m\}$$

$$= P\{S_{m+1} = \gamma(k) \mid S(m) = \gamma(k-1); S_j \notin A, j > m\}$$

$$= P^{\gamma(k-1)}\{S_1 = \gamma(k) \mid S_j \notin A, j > 0\}.$$

This gives the first equality and the second follows from the definition of
$\text{Es}_A(\cdot)$ (see Section 2.2).   □

Another way of stating the above proposition is to say that the loop-
erased walk takes its kth step according to the rule for random walk conditioned not to enter $\{\gamma(0), \ldots, \gamma(k-1)\}$. We now formalize this idea.
Suppose $A \subset Z^d (d \geq 3)$ is a finite set. Then *random walk with (past and
future) taboo set A* is the Markov chain with state space

$$B = \{x \notin A : \text{Es}_A(x) > 0\},$$

and transitions

$$p^A(x, y) = P^x\{S_1 = y \mid \tau_A = \infty\} = \frac{\text{Es}_A(y)}{2d\text{Es}_A(x)}, \quad |y - x| = 1.$$

If we define $q_n^A(x, y)$ by

$$q_n^A(x, y) = P^x\{S_n = y; S_j \notin A, \ j = 0, \ldots n\},$$

then it is easy to check that the n-step transitions for random walk with
taboo set A are given by

$$p_n^A(x, y) = q_n^A(x, y)\frac{\text{Es}_A(y)}{\text{Es}_A(x)}.$$

In particular,

$$p_n^A(x, x) = q_n^A(x, x) \leq p_n = P\{S_n = 0\}. \tag{7.1}$$

By a strong Markov argument identical to that used in deriving (1.19),

$$
\begin{aligned}
P^x\{S_j \neq x, j > 0 \mid \tau_A = \infty\} &= [\sum_{j=0}^{\infty} q_n^A(x, x)]^{-1} \\
&\geq [\sum_{j=0}^{\infty} p_n]^{-1} \\
&= P^x\{S_j \neq x, j > 0\} > 0. \tag{7.2}
\end{aligned}
$$

7.4   Two Dimensions

Since simple random walk is recurrent in two dimensions, we cannot construct the two-dimensional loop-erased walk by erasing loops from an infinite simple random walk. However, we can define the process as a limit of
walks obtained from erasing loops on finite walks. As before, let

$$\xi_m = \inf\{j > 0 : |S_j| \geq m\}.$$

For any $k \leq m$, we can define a measure on Γ_k by taking simple random walks stopped at the random time ξ_m, erasing loops, and considering the first k steps. To be precise, we define $\hat{P}^m = \hat{P}_k^m$ on Γ_k by

$$\hat{P}^m(\gamma) = P\{L[S_0, \ldots, S(\xi_m)](j) = \gamma(j), \; j = 0, \ldots, k\},$$

where L is the loop-erasing operation defined in Section 7.2. This definition is equivalent to that of a Laplacian random walk also cutoff at ξ_m. The proof of the following proposition is identical to that of Proposition 7.3.1.

Proposition 7.4.1 *If $1 \leq k \leq m$, $\gamma_k = [\gamma(0), \ldots, \gamma(k)] \in \Gamma_k$, and $\gamma_{k-1} = [\gamma(0), \ldots, \gamma(k-1)]$, then*

$$
\begin{aligned}
\hat{P}^m(\gamma_k) &= \hat{P}^m(\gamma_{k-1}) P^{\gamma(k-1)}\{S_1 = \gamma(k) \mid \tau_A > \xi_m\} \\
&= \hat{P}^m(\gamma_{k-1}) \frac{P^{\gamma(k)}\{\tau_A > \xi_m\}}{\sum_{y \notin A, |y - \gamma(k-1)| = 1} P^y\{\tau_A > \xi_m\}},
\end{aligned}
$$

where $A = \{\gamma(0), \ldots, \gamma(k-1)\}$ and

$$\tau_A = \inf\{j \geq 1 : S_j \in A\}.$$

We can also give the alternative construction of this walk. Let

$$\tilde{\sigma}_{0,m} = \sup\{j \leq \xi_m : S_j = 0\},$$

and for $1 \leq i \leq m$,

$$\tilde{\sigma}_{i,m} = \sup\{j \leq \xi_m : S_j = S(\tilde{\sigma}_{i-1,m} + 1)\}.$$

Then set $\tilde{S}^m(i) = S(\tilde{\sigma}_{i,m})$ and

$$\hat{P}^m(\gamma) = P\{[\tilde{S}^m(0), \ldots, \tilde{S}^m(k)] = \gamma\}.$$

We would like to define $\hat{P}$ on Γ_k by

$$\hat{P}(\gamma) = \lim_{m \to \infty} \hat{P}^m(\gamma).$$

The next proposition allows us to do so and in the process gives as estimate of the rate of convergence. We call the measure on infinite paths induced by $\hat{P}$ the *two-dimensional loop-erased walk*.

Proposition 7.4.2 *If $n \geq k^2$ and $\gamma \in \Gamma_k$, then for all $m \geq n$,*

$$\hat{P}^m(\gamma) = \hat{P}^n(\gamma)(1 + O(\frac{k^2}{n} \ln \frac{n}{k})).$$

In particular,

$$\lim_{m \to \infty} \hat{P}^m(\gamma) = \hat{P}(\gamma),$$

exists and

$$\hat{P}(\gamma) = \hat{P}^n(\gamma)(1 + O(\frac{k^2}{n} \ln \frac{n}{k})).$$

Proof. Let $\gamma = [\gamma(0), \ldots, \gamma(k)]$ and for $j \leq k$, let $\gamma_j = [\gamma(0), \ldots, \gamma(j)]$ be an initial segment of γ. By Proposition 7.4.1,

$$\hat{P}^m(\gamma_j) = \hat{P}^m(\gamma_{j-1}) \frac{P^{\gamma(j)}\{\tau_A > \xi_m\}}{\sum_{y \notin A, |y - \gamma(j-1)| = 1} P^y\{\tau_A > \xi_m\}}, \tag{7.3}$$

where $A = A_j = \{\gamma(0), \ldots, \gamma(j-1)\}$.

If $y \notin A, |y - \gamma(j-1)| = 1$, then

$$P^y\{\tau_A > \xi_m\} = P^y\{\tau_A > \xi_n\} P^y\{\tau_A > \xi_m \mid \tau_A > \xi_n\}.$$

But,

$$P^y\{\tau_A > \xi_m \mid \tau_A > \xi_n\} = \sum_{z \in \partial C_n} P^y\{S(\xi_n) = z \mid \tau_A > \xi_n\} P^z\{\tau_A > \xi_m\}.$$

By Proposition 2.1.2, since $A \subset C_n$,

$$P^y\{S(\xi_n) = z \mid \tau_A > \xi_n\} = H_{\partial C_n}(0, z)(1 + O(\frac{k}{n} \ln \frac{n}{k}))$$

(actually, Proposition 2.1.2 proves this for $y \in A$, but the identical proof works for $y \in \partial A$). Therefore,

$$P^y\{\tau_A > \xi_m \mid \tau_A > \xi_n\} = P^{\gamma(j)}\{\tau_A > \xi_m \mid \tau_A > \xi_n\}(1 + O(\frac{k}{n} \ln \frac{n}{k})).$$

If we plug this into (7.3) we obtain

$$\begin{aligned} \hat{P}^m(\gamma_j) &= \hat{P}^m(\gamma_{j-1}) \frac{P^{\gamma(j)}\{\tau_A > \xi_n\}}{\sum_{y \in A, |y - \gamma(j-1)| = 1} P^y\{\tau_A > \xi_n\}}(1 + O(\frac{k}{n} \ln \frac{n}{k})) \\ &= \frac{\hat{P}^m(\gamma_{j-1})\hat{P}^n(\gamma_j)}{\hat{P}^n(\gamma_{j-1})}(1 + O(\frac{k}{n} \ln \frac{n}{k})). \end{aligned}$$

If we iterate this for $1 \leq j \leq k$, we get

$$\begin{aligned} \hat{P}^m(\gamma) &= \hat{P}^n(\gamma)(1 + O(\frac{k}{n} \ln \frac{n}{k}))^k \\ &= \hat{P}^n(\gamma)(1 + O(\frac{k^2}{n} \ln \frac{n}{k})). \quad \square \end{aligned}$$

An immediate consequence of this proposition is that for any $\gamma \in \Gamma_k$,

$$\hat{P}(\gamma) = \hat{P}^{k^3}(\gamma)(1 + o(k^{-1/2})). \tag{7.4}$$

7.5 Estimates on Amount Erased

In order to analyze the behavior of $\hat{S}(n)$ for large n, we will need to investigate how many steps of the simple random walk remain after loops have been erased. We first consider the case $d \geq 3$ where the loop-erased walk is constructed by erasing loops from an infinite simple random walk. Recall the definition of σ_i from Section 7.3. An equivalent way of describing σ_i is

$$\sigma(i) = \sigma_i = \sup\{j : S(j) = \hat{S}(i)\}.$$

We define $\rho(j)$ to be the "inverse" of $\sigma(i)$ in the sense

$$\rho(j) = i \text{ if } \sigma_i \leq j < \sigma_{i+1}.$$

Then,

$$\rho(\sigma(i)) = i, \tag{7.5}$$

$$\sigma(\rho(j)) \leq j, \tag{7.6}$$

and

$$\hat{S}(i) = S(\sigma(i)).$$

Let Y_n be the indicator function of the event "the nth point of S is not erased", i.e.,

$$Y_n = \begin{cases} 1 & \text{if } \sigma(i) = n \text{ for some } i \geq 0, \\ 0 & \text{otherwise.} \end{cases}$$

Then,

$$\rho(n) = \sum_{j=1}^{n} Y_j$$

is the number of points remaining of the first n points after loops are erased. We let $a_n = E(Y_n)$ be the probability that the nth point is not erased. Let ω_n be the path obtained by erasing loops on the first n steps of S, i.e.,

$$\omega_n = L[S_0, \ldots, S_n].$$

Then $Y_n = 1$ if and only if

$$\omega_n \cap S[n+1, \infty) = \emptyset,$$

i.e., if the loop-erased walk up to time n and the simple random walk after time n do not intersect. Therefore,

$$P\{Y_n = 1 \mid S_0, \ldots, S_n\} = \text{Es}_{\omega_n}(S_n)$$

and
$$a_n = E[\mathrm{Es}_{\omega_n}(S_n)]. \tag{7.7}$$

By translating the origin this can be restated: if S^1, S^2 are independent simple random walks starting at the origin, then

$$a_n = P\{L^R[S_0^1, \ldots, S_n^1] \cap S^2[1, \infty) = \emptyset\},$$

or equivalently by Proposition 7.2.1,

$$a_n = P\{L[S_0^1, \ldots, S_n^1] \cap S^2[1, \infty) = \emptyset\}.$$

We can extend everything to the $d = 2$ case. Fix n and let $m = n^3$ (we choose $m = n^3$ so that we can use (7.4)). Then, using the notation of the last section, for $i \leq m$,

$$\tilde{\sigma}(i) = \tilde{\sigma}_{i,m} = \sup\{j : S_j = \tilde{S}^m(i)\},$$

and we can define the inverse $\tilde{\rho}$ by

$$\tilde{\rho}(j) = \tilde{\rho}_m(j) = i \text{ if } \tilde{\sigma}(i) \leq j < \tilde{\sigma}(i+1).$$

Then (7.5) and (7.6) hold for $\tilde{\sigma}$ and $\tilde{\rho}$. Define $\tilde{Y}_j = \tilde{Y}_{j,m}$ to be the indicator function of the event "the jth point of S is not erased before ξ_m," i.e.,

$$\tilde{Y}_j = \begin{cases} 1 & \text{if } \tilde{\sigma}(i) = j \text{ for some } i \geq 0, \\ 0 & \text{otherwise.} \end{cases}$$

Then

$$\tilde{\rho}(n) = \sum_{j=1}^{n} \tilde{Y}_j$$

is the number of points remaining after erasing loops through time ξ_m. If $\omega_j = L[S_0, \ldots, S_j]$, then

$$P\{\tilde{Y}_j = 1 \mid S_0, \ldots, S_j\} = P^{S_j}\{\xi_m < \tau_{\omega_j}\}.$$

We will get bounds on the number of points erased by comparing to the number of points remaining afer a finite number of steps of the random walk. This will work for $d = 2$ as well as $d \geq 3$. Fix m and define $\hat{S}^m$ by

$$\hat{S}^m = L[S_0, \ldots, S_m].$$

The length of $\hat{S}^m$ is a random variable. Define

$$\sigma_m(i) = \sigma_{i,m} = \sup\{j \leq m : S_j = \hat{S}^m(i)\},$$

and $\rho_m(i)$ the "inverse" of σ_m by

$$\rho_m(j) = i \text{ if } \sigma_m(i) \leq j < \sigma_m(i+1).$$

Note that $\rho_m(j) < \infty$ for each $j \leq m$ and $\rho_m(m)$ is exactly the length of the path $\hat{S}^m$. As before,

$$\rho_m(\sigma_m(i)) = i, \quad i \leq \rho_m(m), \tag{7.8}$$

$$\sigma_m(\rho_m(j)) \leq j, \quad j \leq m. \tag{7.9}$$

For $n < m$, define $Y_{n,m}$ to be the indicator function of the event "the nth points is not erased by time m," i.e.,

$$Y_{n,m} = \begin{cases} 1 & \text{if } \sigma_m(i) = n \text{ for some } i \geq 0, \\ 0 & \text{otherwise.} \end{cases}$$

Then if $n < m$,

$$\rho_m(n) = \sum_{j=1}^{n} Y_{j,m}.$$

If the nth point is erased by time m, then of course it will be erased eventually, i.e., if $n \leq m$ and $d \geq 3$,

$$Y_{n,m} \geq Y_m. \tag{7.10}$$

Similarly, if $d = 2$ and $j \leq m$,

$$Y_{j,m} \geq \tilde{Y}_{j,m}. \tag{7.11}$$

Hence,

$$\rho_m(n) \geq \rho(n), \quad d \geq 3, \tag{7.12}$$

$$\rho_m(n) \geq \tilde{\rho}_{n^3}(n), \quad d = 2. \tag{7.13}$$

The goal of this section is to derive an upper bound which essentially states that $\rho(n)$ grows no faster than $n(\ln n)^{-1/3}$ for $d = 4$; no faster than $n^{5/6}$ for $d = 3$; and $\tilde{\rho}_{n^3}(n)$ grows no faster than $n^{2/3}$ for $d = 2$. Define b_n by

$$b_n = E(\rho(n)) = \sum_{j=0}^{n} E(Y_j), \quad d \geq 3,$$

$$b_n = E(\rho_{n^3}(n)) = \sum_{j=0}^{n} E(Y_{j,n^3}), \quad d = 2.$$

Theorem 7.5.1 *(a) If $d = 4$,*

$$\limsup_{n \to \infty} \frac{\ln b_n - \ln n}{\ln \ln n} \leq -\frac{1}{3}.$$

(b) If $d = 2, 3$,

$$\limsup_{n \to \infty} \frac{\ln b_n}{\ln n} \leq \frac{d+2}{6}.$$

In the proof of the theorem we will need a lower bound on the probability of returning to the origin while avoiding a given set. Recall that if A is a finite subset of Z^d,

$$q_n^A(x, y) = P^x\{S_n = y; S_j \notin A, j = 0, \ldots, n\}.$$

Lemma 7.5.2 *(a) If $d = 4$, for every $\alpha > 0$ there exists a $c_\alpha > 0$ such that if $0 \notin A$ and $\mathrm{Es}_A(0) \geq (\ln n)^{-1/\alpha}$,*

$$q_{2n}^A(0,0) \geq c_\alpha n^{-2} (\mathrm{Es}_A(0))^2 (\ln n)^{-\alpha}.$$

(b) If $d = 3$, for every $\alpha < \infty$, there exists a $c_\alpha > 0$ such that if $0 \notin A$ and $\mathrm{Es}_A(0) \geq n^{-\alpha}$,

$$q_{2n}^A(0,0) \geq c_\alpha n^{-3/2} (\mathrm{Es}_A(0))^2 (\ln n)^{-3}.$$

(c) If $d = 2$, for every $\alpha < \infty$, there exists a $c_\alpha > 0$ such that if $0 \notin A$ and $P\{\tau_A > \xi_n\} \geq n^{-\alpha}$,

$$q_{2n}^A(0,0) \geq c_\alpha n^{-1} (P\{\tau_A > \xi_n\})^2 (\ln n)^{-2}.$$

Proof. We will prove (a) and (c); (b) can be proved similarly. It suffices in each case to prove the result for n sufficiently large.

For (a), by Lemma 1.5.1,

$$P\{|S_n| \geq n^{1/2} (\ln n)^{\alpha/4}\} \leq c_\alpha \exp\{-(\ln n)^{\alpha/4}\}.$$

Therefore for n sufficiently large (depending on α), if $\mathrm{Es}_A(0) \geq (\ln n)^{-1/\alpha}$,

$$
\begin{aligned}
P\{|S_n| \leq n^{1/2} (\ln n)^{\alpha/4}; S_j \notin A, j = 0, 1, \ldots, n\} \\
\geq \quad \mathrm{Es}_A(0) - P\{|S_n| \geq n^{1/2}(\ln n)^{\alpha/4}\} \\
\geq \quad \frac{1}{2} \mathrm{Es}_A(0).
\end{aligned}
$$

In other words,

$$\sum_{|x| \leq n^{1/2}(\ln n)^{\alpha/4}} q_n^A(0, x) \geq \frac{1}{2} \mathrm{Es}_A(0).$$

But by reversibility of simple random walk,

$$
\begin{aligned}
q_{2n}^A(0,0) \;&\geq\; \sum_{|x|\leq n^{1/2}(\ln n)^{\alpha/4}} q_n^A(0,x)^2 \\
&\geq\; cn^{-2}(\ln n)^{-\alpha}\Big[\sum_{|x|\leq n^{1/2}(\ln n)^{\alpha/4}} q_n^A(0,x)\Big]^2 \\
&\geq\; cn^{-2}(\ln n)^{-\alpha}[\mathrm{Es}_A(0)]^2,
\end{aligned}
$$

for n sufficiently large. The second inequality uses the elementary inequality

$$
\sum_{j=0}^{n} s_j^2 \geq \frac{1}{n+1}\Big(\sum_{j=1}^{n} s_j\Big)^2.
$$

For (c), by Lemma 1.5.1,

$$
P\{|S_n| \geq 2\alpha n^{1/2}(\ln n)\} \leq cn^{-2\alpha}.
$$

Therefore for n sufficiently large, if $P\{\tau_A > \xi_n\} \geq n^{-\alpha}$,

$$
\begin{aligned}
P\{|S_n| \leq 2\alpha n^{1/2}(\ln n); \; S_j \notin A, j = 0,1,\dots,n\} \\
\geq\; P\{\tau_A > \xi_n\} - P\{|S_n| \geq 2\alpha n^{1/2}(\ln n)\} \\
\geq\; \frac{1}{2}P\{\tau_A > \xi_n\},
\end{aligned}
$$

or

$$
\sum_{|x|\leq 2\alpha n^{1/2}(\ln n)} q_n^A(0,x) \geq \frac{1}{2}P\{\tau_A > \xi_n\},
$$

and hence for n sufficiently large,

$$
\begin{aligned}
q_{2n}^A(0,0) \;&\geq\; \sum_{|x|\leq 2\alpha n^{1/2}(\ln n)} q_n^A(0,x)^2 \\
&\geq\; c_\alpha n^{-1}(\ln n)^{-2}[P\{\tau_A > \xi_n\}]^2. \qquad \square
\end{aligned}
$$

Proof of Theorem 7.5.1. Fix n and for $0 \leq j \leq n$ let

$$
X_j = X_{j,2n} = \begin{cases} 0 & \text{if } Y_{j-1,2n} = 0, \\ \sigma_{2n}(i+1) - \sigma_{2n}(i) & \text{if } \sigma_{2n}(i) = j-1. \end{cases}
$$

Then

$$
\sum_{j=1}^{n} X_j \leq 2n,
$$

and hence

$$\sum_{j=1}^{n} E(X_j) \leq 2n. \tag{7.14}$$

Recall that $\omega_j = L[S_0, \ldots, S_j]$. We set

$$Z_j = Es_{\omega_j}(S_j) \text{ if } d = 3, 4,$$

$$Z_j = P^{S_j}\{\tau_{\omega_j} > \xi_{n^3}\} \text{ if } d = 2.$$

For any $0 \leq j \leq n$, let T_j be the nearest neighbor of S_j with $T_j \notin \omega_j$ which maximizes $Es_{\omega_j}(\cdot)$ (or $P^{\cdot}\{\xi_{n^3} < \tau_{\omega_j}\}$ if $d = 2$). If there is more than one such point which obtains the maximum choose one arbitrarily. If each nearest neighbor of S_j is in ω_j, choose T_j arbitrarily. If we set $W_j = Es_{\omega_j}(T_j)$ if $d = 3, 4$ and $W_j = P^{T_j}\{\xi_{n^3} < \tau_{\omega_j}\}$ if $d = 2$, then it is easy to check that

$$W_j \geq Z_j.$$

Then for $0 \leq 2r < n$,

$$
\begin{aligned}
P\{X_j = 2r + 1 \mid S_k, 0 \leq k \leq j\} \\
\geq \quad & P\{S_{j+1} = T_j; S_{j+2r+1} = S_{j+1}; S_k \notin \omega_j, j < k \leq 2n; \\
& \quad S_k \neq S_{j+1}, j + 2r + 1 < k \leq 2n \mid S_k, 0 \leq k \leq j\} \\
\geq \quad & (2d)^{-1} q_{2r}^{\omega_j}(T_j, T_j) \tilde{W}_j,
\end{aligned}
$$

where

$$\tilde{W}_j = Es_{\omega_j \cup \{T_j\}}(T_j), \quad d = 3, 4,$$

$$\tilde{W}_j = P^{T_j}\{\xi_{n^3} < \tau_{\omega_j \cup \{T_j\}}\}, \quad d = 2.$$

If $d = 3, 4$, it follows from (7.2) that

$$\tilde{W}_j \geq cW_j \geq cZ_j.$$

For $d = 2$ one can prove in the same way as (7.2) that if $x \in C_{n^3}, A \subset C_{n^3}, x \notin A$,

$$
\begin{aligned}
P^x\{\tau_x > \xi_{n^3} \mid \tau_A > \xi_{n^3}\} &\geq P^x\{\tau_x > \xi_{n^3}\} \\
&\geq c(\ln n)^{-1}
\end{aligned}
$$

The second inequality follows from Proposition 1.6.7. Therefore,

$$\tilde{W}_j \geq c(\ln n)^{-1} W_j \geq c(\ln n)^{-1} Z_j.$$

Let $d = 4$, $\alpha \in (0, \frac{1}{3})$. By Lemma 7.5.2(a), if $Z_j \geq (\ln n)^{-3}, n^{-1/2} \leq 2r < n$, then

$$q_{2r}^{\omega_j}(T_j, T_j) \geq c_\alpha r^{-2} Z_j^2 (\ln n)^{-\alpha}.$$

Therefore,

$$P\{X_j = 2r + 1 \mid S_k, 0 \leq k \leq j\} \geq c_\alpha r^{-2} Z_j^3 (\ln n)^{-\alpha}.$$

If we sum over $n^{-1/2} \leq 2r < n$, ,

$$E(X_j \mid S_k, 0 \leq k \leq j) \geq c_\alpha (\ln n)^{1-\alpha} Z_j^3 I\{Z_j \geq (\ln n)^{-3}\},$$

and hence,

$$E(X_j) \geq c_\alpha (\ln n)^{1-\alpha} E(Z_j^3 I\{Z_j \geq (\ln n)^{-3}\}).$$

Therefore, by (7.14),

$$\sum_{j=0}^{n} E(Z_j^3 I\{Z_j \geq (\ln n)^{-3}\}) \leq c_\alpha n (\ln n)^{\alpha-1},$$

and hence

$$\sum_{j=0}^{n} E(Z_j^3) \leq c_\alpha n (\ln n)^{\alpha-1}.$$

If $s_0, \ldots, s_n \geq 0$,

$$\sum_{i=0}^{n} s_i^3 \geq (n+1)^{-2} (\sum_{i=0}^{n} s_i)^3.$$

Therefore,

$$\begin{aligned}
\sum_{j=0}^{n} E(Z_j) &\leq (n+1)^{2/3} [\sum_{j=0}^{n} E(Z_j)^3]^{1/3} \\
&\leq (n+1)^{2/3} [\sum_{j=0}^{n} E(Z_j^3)]^{1/3} \\
&\leq c_\alpha n (\ln n)^{(\alpha-1)/3}.
\end{aligned}$$

Since this holds for all $\alpha \in (0, \frac{1}{3})$, we have proved the theorem for $d = 4$.

Similarly, if $d = 2, 3$, by Lemma 7.5.2(b)-(c), if $Z_j \geq n^{-5}$, $n/2 \leq 2r < n$, then

$$q_{2r}^{\omega_j}(T_j, T_j) \geq c r^{-d/2} Z_j^2 (\ln n)^{-d}.$$

If we sum over $n/2 \leq 2r < n$, we get

$$E(X_j) \geq c n^{(4-d)/2} (\ln n)^{-3} E(Z_j^3 I\{Z_j \geq n^{-5}\}),$$

and arguing as before,

$$\sum_{j=0}^{n} E(Z_j) \leq (n+1)^{2/3} [\sum_{j=0}^{n} E(Z_j^3)]^{1/3} \leq cn^{(d+2)/6}(\ln n). \quad \square$$

It is natural to ask is how good the bound in Theorem 7.5.1 is. Let us consider the case $d = 3, 4$. Then

$$Z_n = \mathrm{Es}_{\omega_n}(S_n),$$

$a_n = E(Z_n)$ and $\rho_n = \sum_{j=0}^{n} a_j$. The proof of the theorem gives a way to estimate $E(Z_n^3)$. While the proof only gives a bound for this quantity in one direction, we conjecture that this bound is sharp and that

$$E(Z_n^3) \approx \begin{cases} (\ln n)^{-1}, & d = 4, \\ n^{-1/2}, & d = 3. \end{cases}$$

The proof then proceeds by estimating $E(Z_n)$ by $E(Z_n^3)^{1/3}$. It is quite likely that this bound is not sharp in low dimensions. A similar problem arose in the analysis of intersections of random walks. let

$$\tilde{f}(n) = P\{S[0, n] \cap S[n+1, \infty) = \emptyset\}.$$

Then $\tilde{f}(n) = E(V_n)$ where

$$V_n = \mathrm{Es}_{S[0,n]}(S_n).$$

It is not easy to estimate $E(V_n)$ (a large portion of Chapters 3-5 is devoted to this problem). However, the analysis of two-sided walks (Theorem 3.5.1) allows us to show that

$$E(V_n^2) \approx \begin{cases} (\ln n)^{-1}, & d = 4, \\ n^{-1/2}, & d = 3. \end{cases}$$

For this problem the second moment is relatively easy to estimate while for the loop-erased walk it is the third moment. How much do we lose when we estimate $E(V_n)$ by $E(V_n^2)^{1/2}$? If $d = 4$, we lose very little since by Theorem 4.4.1, $\tilde{f}(n) \approx (\ln n)^{-1/2}$. By analogy we conjecture that we lose little in estimating $E(Z_n)$ by $E(Z_n^3)^{1/3}$ in four dimensions, i.e., we conjecture

$$a_n \approx (\ln n)^{-1/3}, \quad d = 4.$$

For $d = 3$, we expect that the estimate $E(V_n) \leq E(V_n^2)^{1/2}$ is not sharp; in fact $\tilde{f}(n) \approx n^{-\zeta}$ where it is conjectured that $\zeta \in (.28, .29)$. Again by analogy we expect that the estimate for $E(Z_n)$ is not sharp in three dimensions and that $a_n \approx n^{-\alpha}$ for some $\alpha > \frac{1}{6}$. We also do not expect that the estimate in Theorem 7.5.1 will be sharp in two dimensions. Therefore, the estimates for the mean square displacement given in the next section are not conjectured to be sharp. Monte Carlo simulations [31] are consistent with this belief.

7.6 Growth Rate in Low Dimensions

As a corollary to Theorem 7.5.1 we will prove that the mean-square displacement of the loop-erased walk, $\langle |\gamma(n)|^2 \rangle_{\hat{P}} = E(|\hat{S}(n)|^2)$, grows at least as fast as the Flory predictions for the usual self-avoiding walk, i.e., $E(|\hat{S}(n)|^2)$ grows no slower than $n^{6/(2+d)}$ in two and three dimensions. As mentioned in the previous section, it is probably true that the displacement is even greater. Monte Carlo simulations [31] predict

$$E(|\hat{S}(n)|^2) \approx \begin{cases} n^{8/5}, & d = 2, \\ n^{1.23\cdots}, & d = 3. \end{cases}$$

Most of the work in proving the estimate was done in Theorem 7.5.1. We first state an easy lemma about the minimum displacement of simple random walk.

Lemma 7.6.1 (a) If $d \geq 3$, for every $\epsilon > 0$,

$$\lim_{n \to \infty} P\{\inf_{j \geq n} |S_j|^2 \leq n^{1-2\epsilon}\} = 0.$$

(b) If $d = 2$ and

$$D_n = \inf\{|S_j|^2 : n \leq j \leq \xi_{n^3}\},$$

then for every $\epsilon > 0$,

$$\lim_{n \to \infty} P\{D_n \leq n^{1-2\epsilon}\} = \frac{5}{5 + 2\epsilon}.$$

Proof. By the central limit theorem

$$P\{|S_n|^2 \leq n^{1-\epsilon}\} \to 0. \tag{7.15}$$

If $d \geq 3$, by Proposition 1.5.10, if $|x|^2 \geq n^{1-\epsilon}$,

$$P^x\{\inf_{j \geq 0} |S_j|^2 \leq n^{1-2\epsilon}\} \leq O(n^{-\epsilon/2}).$$

This gives (a). For (b), let

$$\sigma = \inf\{k \geq n : |S_k| \geq n^3 \text{ or } |S_k|^2 \leq n^{1-2\epsilon}\}.$$

By the optional sampling theorem, if a is the potential kernel defined in Section 1.6,

$$E(a(S_\sigma)) = E(a(S_n)).$$

By Theorem 1.6.2 it is easy to show that

$$E(a(S_n)) = \frac{1}{\pi} \ln n + O(1).$$

Similarly, using (7.15),

$$E(a(S_\sigma) \mid |S_\sigma|^2 \le n^{1-2\epsilon}) = \frac{1}{\pi}(1 - 2\epsilon) \ln n + O(1),$$

$$E(a(S_\sigma) \| |S_\sigma| \ge n^3) = \frac{6}{\pi} \ln n + O(1).$$

and hence

$$\lim_{n\to\infty} P\{|S_\sigma|^2 \le n^{1-2\epsilon}\} = \frac{5}{5+2\epsilon}. \quad \square$$

Theorem 7.6.2 *If $\hat{S}$ is the loop-erased walk, then for $d = 2, 3$,*

$$\liminf_{n\to\infty} \frac{\ln E(|\hat{S}(n)|^2)}{\ln n} \ge \frac{6}{2+d}.$$

Proof. Let $d = 3$. By Theorem 7.5.1(b), for every $\epsilon > 0$, if n is sufficiently large,

$$E(\rho(n)) \le n^{\frac{5}{6}+\frac{\epsilon}{2}},$$

and hence for every $\epsilon > 0$,

$$P\{\rho(n) \ge n^{\frac{5}{6}+\epsilon}\} \to 0.$$

Note that if $\sigma(j) \le j^{\frac{6}{5}-\epsilon}$,

$$\rho([j^{\frac{6}{5}-\epsilon}]) \ge \rho(\sigma(j)) = j.$$

Therefore, for every $\epsilon > 0$,

$$P\{\sigma(j) \le j^{\frac{6}{5}-\epsilon}\} \to 0.$$

But $\hat{S}(j) = S(\sigma(j))$ and using Lemma 7.6.1, if

$$D_j = \inf\{|S_k|^2 : j^{\frac{6}{5}-\epsilon} \le k < \infty\},$$

then

$$P\{|\hat{S}(j)|^2 \le j^{\frac{6}{5}-2\epsilon}\} \le P\{|D_j| \le j^{\frac{6}{5}-2\epsilon}\} + o_\epsilon(1) \to 0.$$

This gives the result for $d = 3$.

For $d = 2$ it suffices by (7.4) to prove the result for $\tilde{S}^{n^3}$. Then as above we can derive from Theorem 7.5.1 that for every $\epsilon > 0$,

$$P\{\tilde{\sigma}_{j^3}(j) \le j^{\frac{3}{2}-\epsilon}\} \to 0.$$

Therefore, if

$$D_j = \inf\{|S_k|^2 : j^{\frac{3}{2}-2\epsilon} \le k \le \xi_{j^3}\},$$

then

$$P\{|\tilde{S}^{j^3}(j)|^2 \le j^{\frac{3}{2}-2\epsilon}\} \le P\{D_j \le j^{\frac{3}{2}-2\epsilon}\} + o_\epsilon(1) \to c_\epsilon < 1,$$

and hence

$$\liminf_{j\to\infty} P\{|\tilde{S}^{j^3}(j)|^2 \ge j^{\frac{3}{2}-2\epsilon}\} > 0. \quad \square$$

7.7 High Dimensions

We will show that the loop-erased walk appropriately scaled approaches a Brownian motion if $d \ge 4$. If $d \ge 5$, the scaling will just be a constant times the usual scaling for simple random walk, while for $d = 4$ a logarithmic correction term will appear. The key step to proving such convergence is to show that the loop-erasing process is uniform on paths, i.e., that

$$r_n^{-1}\rho(n) \to 1,$$

for some $r_n \to \infty$.

We first consider the case $d \ge 5$. Here it will be convenient to extend S to a two-sided walk. Let S^1 be a simple random walk independent of S and let $S_j, -\infty < j < \infty$, be defined by

$$S_j = \begin{cases} S_j, & 0 \le j < \infty, \\ S_j^1, & -\infty < j \le 0. \end{cases}$$

We call a time j *loop-free* for S if $S(-\infty, j] \cap S(j, \infty) = \emptyset$. By(3.2), for each j,

$$P\{j \text{ loop-free}\} = P\{S(-\infty, 0] \cap S(0, \infty) = \emptyset\} = b > 0.$$

Lemma 7.7.1 *If $d \ge 5$, with probability one, $S(-\infty, \infty)$ has infinitely many positive loop-free points and infinitely many negative loop-free points.*

Proof. Let X be the number of positive loop-free points. We call a time j *n-loop-free* if $S[j-n, j] \cap S(j, j+n] = \emptyset$. Then

$$P\{j \text{ n-loop-free}\} = P\{S[-n, 0] \cap S(0, n] = \emptyset\} = b_n,$$

and $b_n \to b$. Let $V_{i,n}$ be the event $\{(2i-1)n$ is loop-free$\}$ and $W_{i,n}$ the event $\{(2i-1)n$ is n-loop-free$\}$. Note that for a given n, the events $W_{i,n}$, $i = 1, 2, \ldots$, are independent. For any $k < \infty$, $\epsilon > 0$, find m such that if Y is a binomial random variable with parameters m and ϵ, $P\{Y < k\} \leq \epsilon$. Then

$$
\begin{aligned}
P\{X \geq k\} \ &\geq\ P\{\sum_{i=1}^{m} I(V_{i,n}) \geq k\} \\
&\geq\ P\{\sum_{i=1}^{m} I(W_{i,n}) \geq k\} - P\{\sum_{i=1}^{m} I(W_{i,n} \setminus V_{i,n}) \geq 1\} \\
&\geq\ 1 - \epsilon - m(b - b_n).
\end{aligned}
$$

Now choose n so that $m(b - b_n) \leq \epsilon$. Then $P\{X \geq k\} \geq 1 - 2\epsilon$. Since this holds for all $k < \infty, \epsilon > 0$, we must have $P\{X = \infty\} = 1$. A similar proof shows that the number of negative loop-free points is infinite with probability one. $\Box$

Theorem 7.7.2 *If $d \geq 5$, there exists an $a > 0$ such that with probability one*

$$
\lim_{n \to \infty} \frac{\rho(n)}{n} = a.
$$

Proof: Order the loop-free points of $S(-\infty, \infty)$,

$$
\cdots \leq j_{-2} \leq j_{-1} \leq j_0 \leq j_1 \leq j_2 \leq \cdots,
$$

with

$$
j_0 = \inf\{j \geq 0 : j \text{ loop-free}\}.
$$

We can erase loops on the two-sided path $S(-\infty, \infty)$ by erasing separately on each piece $S[j_i, j_{i+1}]$. Let $\tilde{Y}_n$ be the indicator function of the event "the nth point is not erased in this procedure," i.e., $\tilde{Y}_n = 1$ if and only if $j_i \leq n < j_{i+1}$ for some i and

$$
L(S[j_i, n]) \cap S(n, j_{i+1}] = \emptyset.
$$

We note that the $\tilde{Y}_n$ form a stationary, ergodic sequence. Therefore by a standard ergodic theorem (see [9], Theorem 6.28), with probability one,

$$
\lim_{n \to \infty} \frac{1}{n} \sum_{j=0}^{n} \tilde{Y}_j = E(\tilde{Y}_0).
$$

If instead we erase loops only on the path $S[0, \infty)$, ignoring $S(-\infty, 0)$, the self-avoiding path we get may be slightly different. However, it is easy to see that if $n \geq j_0$, then $Y_n = \tilde{Y}_n$, where Y_n is as defined in section 5. Therefore, since $j_0 < \infty$, with probability one,

$$\lim_{n \to \infty} \frac{\rho(n)}{n} = \lim_{n \to \infty} \frac{1}{n} \sum_{j=1}^{\infty} Y_n = E(\tilde{Y}_0) \doteq a.$$

To see that $a > 0$ we need only note that

$$a \geq P\{0 \text{ loop-free}\} > 0. \quad \square$$

We cannot use such a proof for $d = 4$ since $S(-\infty, \infty)$ contains no (two-sided) loop-free points. However, we will be able to make use of one-sided loop-free points. Let $I_n = I(n)$ be the indicator function of the event "n is a (one-sided) loop-free point," i.e.,

$$S[0, n] \cap S(n, \infty) = \emptyset.$$

The first lemma shows that the property of being loop-free is in some sense a local property.

Lemma 7.7.3 *Let $d = 4$ and*

$$U_n = \{S[0, n] \cap S(n, \infty) = \emptyset\}$$

$$V_{n,k} = \{S[k - n(\ln n)^{-9}, k] \cap S(k, k + n(\ln n)^{-9}] = \emptyset\}.$$

Then for all k with $n(\ln n)^{-9} \leq k \leq n$,

$$P(V_{n,k}) = P(U_n)(1 + O(\frac{\ln \ln n}{\ln n})).$$

Proof. It suffices to prove the lemma for $k = n$. We write U for U_n and V for $V_{n,n}$. Let

$$\overline{V} = \overline{V}_n = \{S[n - n(\ln n)^{-9}, n] \cap S(n, \infty) = \emptyset\}.$$

Then by Proposition 4.4.4 and Theorem 4.3.6,

$$P(\overline{V}) = P(U)(1 + O(\frac{\ln \ln n}{\ln n})). \tag{7.16}$$

Let

$$W = W_n = \{S[n - n(\ln n)^{-18}, n] \cap S(n, n + n(\ln n)^{-9}] = \emptyset\},$$

$$\overline{W} = \overline{W}_n = \{S[n - n(\ln n)^{-18}, n] \cap S(n, \infty) = \emptyset\}.$$

Then again by Proposition 4.4.4 and Theorem 4.3.6,

$$P(\overline{W}) = P(\overline{V})(1 + O(\frac{\ln \ln n}{\ln n})). \qquad (7.17)$$

But by (3.9) and Proposition 4.3.1(iv),

$$P(W \setminus \overline{W}) \leq P\{S[n - n(\ln n)^{-18}, n] \cap S[n + n(\ln n)^{-9}, \infty) = \emptyset\}$$
$$= o((\ln n)^{-2}).$$

Since $P(\overline{W}) \approx (\ln n)^{-1/2}$ (Theorem 4.4.1), this implies

$$P(W) = P(\overline{W})(1 + O(\frac{\ln \ln n}{\ln n})). \qquad (7.18)$$

But $\overline{V} \subset V \subset W$, so (7.16) - (7.18) imply

$$P(V) = P(U)(1 + O(\frac{\ln \ln n}{\ln n})). \quad \square$$

The next lemma will show that there are a lot of loop-free points on a path. Suppose $0 \leq j < k < \infty$, and let $Z(j, k)$ be the indicator function of the event "there is no loop-free point between j and k," i.e.,

$$\{I_m = 0, j \leq m \leq k\}.$$

Then by Theorem 4.3.6(ii), if $d = 4$,

$$E(Z(n - n(\ln n)^{-6}, n)) \geq P\{S[0, n - n(\ln n)^{-6}] \cap S(n + 1, \infty) \neq \emptyset\}$$
$$\geq c\frac{\ln \ln n}{\ln n}.$$

The next lemma gives a similar bound in the opposite direction.

Lemma 7.7.4 If $d = 4$, for any n and k with $n(\ln n)^{-6} \leq k \leq n$,

$$E(Z(k - n(\ln n)^{-6}, k)) \leq c\frac{\ln \ln n}{\ln n}.$$

Proof. It suffices to prove the result for $k = n$. Fix n; let $m = m_n = [(\ln n)^2]$; and choose $j_1 < j_2 < \ldots < j_m$ (depending on n) satisfying

$$n - n(\ln n)^{-6} \leq j_i \leq n, \quad i = 1, \ldots, m,$$

$$j_i - j_{i-1} \geq 2n(\ln n)^{-9}, \quad i = 2, \ldots, m.$$

Let $J(k, n)$ be the indicator function of

$$\{S[k - n(\ln n)^{-9}, k] \cap S(k, k + n(\ln n)^{-9}] = \emptyset\},$$

and

$$X = X_n = \sum_{i=1}^{m} I(j_i),$$

$$\overline{X} = \overline{X}_n = \sum_{i=1}^{m} J(j_i, n).$$

By Lemma 7.7.3,

$$E(J(j_i, n)) = E(I(j_i))(1 + O(\frac{\ln \ln n}{\ln n})),$$

and hence

$$E(\overline{X}) = E(X)(1 + O(\frac{\ln \ln n}{\ln n})),$$

$$E(\overline{X} - X) \le c\frac{\ln \ln n}{\ln n}E(X). \tag{7.19}$$

Note that

$$E(Z(n - n(\ln n)^{-6}, n)) \le P\{X = 0\}$$
$$\le P\{\overline{X} - X \ge \frac{1}{2}E(X)\} + P\{\overline{X} \le \frac{1}{2}E(X)\}.$$

The first term is estimated easily using (7.19),

$$P\{\overline{X} - X \ge \frac{1}{2}E(X)\} \le 2[E(X)]^{-1}E(\overline{X} - X)$$
$$\le c\frac{\ln \ln n}{\ln n}.$$

To estimate the second term, note that $J(j_1, n), \ldots, J(j_m, n)$ are independent and hence

$$\mathrm{Var}(\overline{X}) = \sum_{i=1}^{m} \mathrm{Var}(J(j_i, n)) \le \sum_{i=1}^{m} E(J(j_i, n)) \le E(\overline{X}),$$

and hence by Chebyshev's inequality, for n sufficiently large,

$$P\{\overline{X} \le \frac{1}{2}E(X)\} \le P\{\overline{X} \le \frac{2}{3}E(\overline{X})\} \le c[E(\overline{X})]^{-1}.$$

But by Theorem 4.4.1,

$$E(\overline{X}) \ge c(\ln n)^2 E(I(n)) \ge c(\ln n)^{11/8}.$$

Hence, $P\{\overline{X} \le \frac{1}{2}E(X)\} \le c(\ln n)^{-11/8}$ and the lemma is proved. $\square$

Recall from Section 7.5 that Y_n is the indicator function of the event "the nth point is not erased" and $a_n = E(Y_n)$. Suppose that for some $0 \le k \le n$, loops are erased only on $S[k, \infty)$, so that S_k is considered to be the origin. Let $Y_{n,k}$ be the probability that S_n is erased in this procedure. Clearly $E(Y_{n,k}) = a_{n-k}$. Now suppose $0 \le k \le n - n(\ln n)^{-6}$ and $Z(n - n(\ln n)^{-6}, n) = 0$, i.e., that there exists a loop-free point between $n - n(\ln n)^{-6}$ and n. Then it is easy to check that $Y_{n,k} = Y_n$, and hence by the previous lemma,

$$P\{Y_n \ne Y_{n,k}\} \le E(Z(n - n(\ln n)^{-6}, n)) \le c \frac{\ln \ln n}{\ln n}.$$

Therefore, for $n(\ln n)^{-6} \le k \le n$,

$$|a_k - a_n| \le c a_n (\ln n)^{-3/8},$$

i.e.,

$$a_k = a_n(1 + o((\ln n)^{-1/4})). \tag{7.20}$$

The second inequality follows from the estimate $a_n \ge f(n) \approx (\ln n)^{-1/2}$. We can combine this with Theorem 7.5.1(a) to conclude

$$-\frac{1}{2} \le \liminf_{n \to \infty} \frac{\ln a_n}{\ln \ln n} \le \limsup_{n \to \infty} \frac{\ln a_n}{\ln \ln n} \le -\frac{1}{3}.$$

We also conclude

$$E(\rho(n)) \sim n a_n. \tag{7.21}$$

The following theorem shows that the number of points remaining after erasing loops satisfies a weak law of large numbers.

Theorem 7.7.5 *If* $d = 4$,

$$\frac{\rho(n)}{n a_n} \to 1,$$

in probability.

Proof. For each n, choose

$$0 \le j_0 < j_1 < \cdots < j_m = n$$

such that $(j_i - j_{i-1}) \sim n(\ln n)^{-2}$, uniformly in i. Then $m \sim (\ln n)^2$. Erase loops on each interval $[j_i, j_{i+1}]$ separately (i.e., do finite loop-erasing on $S[j_i, j_{i+1}]$.) Let $\tilde{Y}_k$ be the indicator function of the event "S_k is not erased in this finite loop-erasing." Let $K_0 = [0, 0]$, and for $i = 1, \ldots, m$, let K_i be the interval

$$K_i = [j_i - n(\ln n)^{-6}, j_i].$$

Let $R_i, i = 1, \ldots, m$, be the indicator function of the complement of the event, "there exist loop-free points in both K_{i-1} and K_i," i.e., the complement of the event

$$\{Z(j_{i-1} - n(\ln n)^{-6}, j_{i-1}) = Z(j_i - n(\ln n)^{-6}, j_i) = 0\}.$$

By Lemma 7.7.4,

$$E(R_i) \leq c \frac{\ln \ln n}{\ln n}. \tag{7.22}$$

Note that if $j_i \leq k < j_{i+1} - n(\ln n)^{-6}$ and $R_i = 0$, then

$$Y_k = \tilde{Y}_k.$$

Therefore, for n sufficiently large,

$$
\begin{aligned}
|\sum_{k=0}^{n} Y_k - \sum_{k=0}^{n} \tilde{Y}_k| &\leq c[m(n(\ln n)^{-6}) + 2n(\ln n)^{-2} \sum_{i=1}^{m} R_i] \\
&\leq cn(\ln n)^{-4} + cn(\ln n)^{-2} \sum_{i=1}^{m} R_i.
\end{aligned}
$$

But by (7.22),

$$
\begin{aligned}
P\{\sum_{i=1}^{m} R_i \geq (\ln n)^{5/4}\} &\leq (\ln n)^{-5/4} E(\sum_{i=1}^{m} R_i) \\
&\leq c \ln \ln n (\ln n)^{-1/4}.
\end{aligned}
$$

Therefore,

$$P\{|\sum_{k=0}^{n} Y_k - \sum_{k=0}^{n} \tilde{Y}_k| \geq cn(\ln n)^{-3/4}\} \to 0.$$

Since $na_n \geq cn(\ln n)^{-5/8}$, this implies

$$(na_n)^{-1}(\sum_{k=0}^{n} Y_k - \sum_{k=0}^{n} \tilde{Y}_k) \to 0 \tag{7.23}$$

in probability. We can write

$$\sum_{k=0}^{n} \tilde{Y}_k = 1 + \sum_{i=1}^{m} X_i,$$

where $X_1, \ldots, X_m$ are the independent random variables,

$$X_i = \sum_{k=j_{i-1}}^{j_i - 1} \tilde{Y}_k.$$

Note that

$$\text{Var}(X_i) \le E(X_i^2) \le \|X_i\|_\infty E(X_i) \le cn(\ln n)^{-2} E(X_i),$$

and hence

$$\text{Var}(\sum_{k=0}^{n} \tilde{Y}_k) \le cn(\ln n)^{-2} E(\sum_{k=0}^{n} \tilde{Y}_k).$$

Therefore, by Chebyshev's inequality,

$$P\{|\sum_{k=0}^{n} \tilde{Y}_k - E(\sum_{k=0}^{n} \tilde{Y}_k)| \ge (\ln n)^{-1/2} E(\sum_{k=0}^{n} \tilde{Y}_k)\}$$

$$\le \quad cn(\ln n)^{-1} [E(\sum_{k=0}^{n} \tilde{Y}_k)]^{-1}$$

$$\le \quad c(\ln n)^{-3/8}.$$

This implies

$$[E(\sum_{k=0}^{n} \tilde{Y}_k)]^{-1} \sum_{k=0}^{n} \tilde{Y}_k \to 1$$

in probability. It is easy to check, using (7.20) that $E(\sum_{k=0}^{n} \tilde{Y}_k) \sim na_n$ and hence by (7.23),

$$(na_n)^{-1} \rho(n) = (na_n)^{-1} \sum_{k=0}^{n} Y_k \to 1$$

in probability.   □
   We finish this section by showing that the loop-erased walk converges to Brownian motion if $d \ge 4$. This is essentially a consequence of Theorems 7.7.2 and 7.7.5. Recall from Section 7.5 that

$$\hat{S}(n) = S(\sigma(n)),$$

where σ is the "inverse" of ρ. If $d \ge 5$, by Theorem 7.7.2, with probability one,

$$\lim_{n\to\infty} \frac{\rho(\sigma(n))}{\sigma(n)} = a,$$

and hence by (7.5),

$$\lim_{n\to\infty} \frac{\sigma(n)}{n} = \frac{1}{a}. \tag{7.24}$$

For $d = 4$, since $a_n \geq c(\ln n)^{-5/8}$, it follows from (7.20) that $a_{[n/a_n]} \sim a_n$. Therefore, by Theorem 7.7.5,

$$\frac{\rho([n/a_n])}{n} \to 1$$

in probability. It is not hard then using the monotonicity of ρ to show that

$$\frac{\sigma(n)a_n}{n} \to 1 \tag{7.25}$$

in probability.

We will use $\Rightarrow$ to denote weak convergence in the metric space $C[0, 1]$ with the sup norm. Then the standard invariance principle states that if $W_n(t) = dn^{-1/2}S([nt])$, then $W_n(t) \Rightarrow B(t)$, where B is a standard Brownian motion in R^d. Suppose $b_n \to \infty$ and

$$r_n(t) \doteq \frac{\sigma([nt])}{b_n} \Rightarrow t.$$

Then by the continuity of Brownian motion (or more precisely, the tightness in $C[0, 1]$ of the sequence W_n),

$$\frac{S(\sigma([nt]))}{\sqrt{b_n}} - \frac{S([b_n t])}{\sqrt{b_n}} \Rightarrow 0,$$

and hence

$$\frac{dS(\sigma([nt]))}{\sqrt{b_n}} \Rightarrow B(t).$$

If $d \geq 5$, it follows immediately from (7.24) that $n^{-1}\sigma([nt])a \Rightarrow t$. For $d = 4$, we need to be a little careful. Fix $\epsilon > 0$ and choose $k \geq 3\epsilon^{-1}$. Let $\delta > 0$. Then by (7.25), for all n sufficiently large,

$$P\{|\frac{\sigma([nj/k])a_{[nj/k]}}{[nj/k]} - 1| \geq \frac{\epsilon}{4}\} \leq \frac{\delta}{k}, \quad j = 1, \ldots, k.$$

Since $a_{[n/k]} \sim a_n$ (see (7.20)) this implies for n sufficiently large

$$P\{|\frac{\sigma([nj/k])a_n}{n} - \frac{j}{k}| \geq \frac{\epsilon}{2}\} \leq \frac{\delta}{k}, \quad j = 1, \ldots, k.$$

But since σ is increasing and $k \geq 3\epsilon^{-1}$, this implies for n sufficiently large,

$$P\{\sup_{0 \leq t \leq 1} |\frac{\sigma([nt])a_n}{n} - t| \geq \epsilon\} \leq \delta.$$

Since this holds for any $\epsilon, \delta > 0$,

$$\frac{\sigma([nt])a_n}{n} \Rightarrow t, \quad d = 4.$$

We have therefore proved the following.

Theorem 7.7.6 *(a) If $d \geq 5$, and*

$$\hat{W}_n(t) = \frac{d\sqrt{a}\hat{S}([nt])}{\sqrt{n}},$$

then $\hat{W}_n(t) \Rightarrow B(t)$, where B is a standard Brownian motion.
    (b) If $d = 4$, and

$$\hat{W}_n(t) = \frac{d\sqrt{a_n}\hat{S}([nt])}{\sqrt{n}},$$

then $\hat{W}_n(t) \Rightarrow B(t)$, where B is a standard Brownian motion.

Bibliography

[1] Ahlfors, L. (1973). *Conformal Invariance. Topics in Geometric Function Theory.* McGraw-Hill.

[2] Aizenman, M. (1985). The intersection of Brownian paths as a case study of a renormalization method for quantum field theory. Commun. Math. Phys. **97** 111-124.

[3] Amit, D.J., G. Parisi, and L. Peliti (1983). Asymptotic behavior of the "true" self-avoiding walk. Phys. Rev. B **27** 1635-1645.

[4] Berg, P. and MacGregor, J. (1966). *Elementary Partial Differential Equations.* Holden-Day.

[5] Berretti, A. and A. Sokal (1985). New Monte Carlo method for the self-avoiding walk. J. Stat. Phys. **40** 483-531.

[6] Beyer, W. A. and M. B. Wells (1972). Lower bound for the connective constant of a self-avoiding walk on a square lattice. J. of Comb. Theor. **13** 176-182.

[7] Billingsley, P. (1986). *Probability and Measure.* John Wiley & Sons.

[8] Brandt, A. (1966). Estimates for difference quotients of solutions of Poisson type difference equations. Math. Comp. **20** 473-499.

[9] Breiman, L. (1968). *Probability.* Addison-Wesley.

[10] Brydges, D. and Spencer, T. (1985). Self-avoiding walk in 5 or more dimensions. Commun. Math. Phys. **97** 125-148.

[11] Burdzy, K. and G. Lawler (1990). Non-intersection exponents for random walk and Brownian motion. Part I: Existence and an invariance principle. Prob. Theor. and Rel. Fields **84** 393-410.

[12] —— (1990). Non-intersection exponents for random walk and Brownian motion. Part II: Estimates and applications to a random fractal. Annals of Prob. **18** 981-1009.

[13] Burdzy, K., G. Lawler, and T. Polaski (1989). On the critical exponent for random walk intersections. J. Stat. Phys. **56** 1-12.

[14] Deuschel, J.-D., and D. Stroock (1989). *Large Deviations*. Academic Press.

[15] Dubins, L., A. Orlitsky, J. Reeds, and L. Shepp (1988). Self-avoiding random loops. IEEE Trans. Inform. Theory **34** 1509-1516.

[16] Duplantier, B. (1987). Intersections of random walks: a direct renormalization approach. Commun. Math. Phys. **117** 279-330.

[17] Duplantier, B. and K.-H. Kwon (1988). Conformal invariance and intersections of random walks. Phys. Rev. Lett. **61** 2514-1517.

[18] Durrett, R. (1984). *Brownian Motion and Martingales in Analysis*. Wadsworth.

[19] Dvoretsky, A., P. Erdös, and S. Kakutani (1950). Double points of paths of Brownian motions in n-space. Acta. Sci. Math. Szeged **12** 75-81.

[20] Edwards, S. F. (1965). The statistical mechanics of polymers with excluded volume. Proc. Phys. Sci. **85** 613-624.

[21] Ellis, R. (1985). *Entropy, Large Deviations, and Statistical Mechanics*. Springer-Verlag.

[22] Erdös, P. and S. J. Taylor (1960). Some intersection properties of random walk paths. Acta. Math. Sci. Hung. **11** 231-248.

[23] Felder, G. and J. Fröhlich (1985). Intersection properties of simple random walks: a renormalization group approach. Commun. Math. Phys. **97** 111-124.

[24] Feller, W. (1968). *An Introduction to Probability Theory and its Applications, Vol. I*. John Wiley & Sons.

[25] —— (1971). *An Introduction to Probability Theory and its Applications, Vol. II*. John Wiley & Sons.

[26] Flory, P. (1949). The configuration of real polymer chain. J. Chem. Phys. **17** 303-310.

[27] de Gennes, P-G (1979). *Scaling Concepts in Polymer Physics*. Cornell Univesity Press.

[28] Gilbarg, D. and N. S. Trudinger (1983), *Elliptic Partial Differential Equations of Second Order*. Springer-Verlag.

[29] Guttmann, A. (1978). On the zero-field susceptibility in the $d = 4, n = 0$ limit: analyzing for confluent logarithmic singularites. J. Phys. A. **11** L103-L106.

[30] —— (1987). On the critical behavior of self-avoiding walks. J. Phys A. **20** 1839-1854.

[31] Guttmann, A. and R. Bursill (1990). Critical exponent for the loop erased self-avoiding walk by Monte Carlo methods. J. Stat. Phys. **59** 1-9.

[32] Hammersley, J. M. (1961). The number of polygons on a lattice. Proc. Camb. Phil. Soc. **57** 516-523.

[33] Kesten, H. (1962). On the number of self-avoiding walks. J. Math. Phys. **4** 960-969.

[34] —— (1964). On the number of self-avoiding walks. II. J. Math. Phys. **5** 1128-1137.

[35] —— (1987). Hitting probabilities of random walks on Z^d. Stoc. Proc. and Appl. **25** 165-184.

[36] —— (1987). How long are the arms in DLA? J. Phys. A. **20** L29-L33.

[37] —— (1990). Upper bounds for the growth rate of DLA. Physica A **168** 529-535.

[38] Lawler, G. (1980). A self-avoiding random walk. Duke. Math. J. **47** 655-694.

[39] —— (1982). The probability of intersection of independent random walks in four dimensions. Commun. Math. Phys. **86** 539-554.

[40] —— (1985). Intersections of random walks in four dimensions II. Commun. Math. Phys. **97** 583-594.

[41] —— (1985). The probability of intersection of three random walks in three dimensions. Unpublished manuscript.

[42] —— (1986). Gaussian behavior of loop-erased self-avoiding random walk in four dimensions. Duke Math. J. **53** 249-270.

[43] —— (1988). Loop-erased self-avoiding random walk in two and three dimensions. J. Stat. Phys. **50** 91-108.

[44] —— (1989). The infinite self-avoiding walk in high dimensions. Annals of Prob. **17** 1367-1376.

[45] —— (1989). Intersections of random walks with random sets. Israel J. Math. **65** 113-132.

[46] Lawler, G. and A. Sokal (1988). Bounds on the L^2 spectrum for Markov chains and Markov processes: a generalization of Cheeger's inequality. Trans. AMS **309** 557-580.

[47] Le Gall, J.-F. (1986). Propriétés d'intersection des marches aléatoires. I. Convergence vers le temps local d'intersection. Commun. Math. Phys. **104** 471-507.

[48] Le Guillou, J. C. and Zinn-Justin, J. (1989). Accurate critical exponents from field theory. J. Phys. France **50** 1365-1370.

[49] Li, B. and A. Sokal (1990). High-precision Monte Carlo test of the conformal-invariance predictions for two-dimensional mutually avoiding walks. J. Stat Phys. **61** 723-748.

[50] Lyklema, J. W. , C. Evertsz and L. Pietronero (1986). The Laplacian random walk. Europhys. Lett. **2** 77-82.

[51] Madras, N. (1988). End patterns of self-avoiding walks. J. Stat. Phys. **53** 689-701.

[52] —— (1991). Bounds on the critical exponent of self-avoiding polygons. Festschrift in honor of Frank Spitzer (R. Durrett and H. Kesten, ed.). Birkhäuser-Boston.

[53] Madras, N., A. Orlitsky, and L. Shepp (1990). Monte Carlo generation of self-avoiding walks with fixed endpoints and fixed lengths. J. Stat. Phys. **58** 159-183.

[54] Madras, N. and A. Sokal (1988). The pivot algorithm: a highly efficient Monte Carlo method for the self-avoiding walk. J. Stat. Phys. **50** 109-186.

[55] Mandelbrot, B. (1983). *The Fractal Geometry of Nature*. W. H. Freeman.

[56] Nienhuis, B. (1982). Exact critical exponents of $O(n)$ models in two dimensions. Phys. Rev. Lett. **49** 1062-1065.

[57] —— (1984). Critical behavior of two-dimensional spin models and charge asymmetry in the Coulomb gas. J. Stat. Phys. **34** 731-761.

[58] Park, Y. (1989). Direct estimates on intersection probabilities of random walks. J. Stat. Phys. **57** 319-331.

[59] Polaski, T. (1991). Ph.D. Dissertation, Duke University.

[60] Port, S. and C. Stone (1978). *Brownian Motion and Classical Potential Theory*. Academic Press.

[61] Scott, D. (1990). A non-integral-dimensional random walk. J. Theor. Prob. **3** 1-7.

[62] Slade, G. (1987). The diffusion of self-avoiding random walk in high dimensions. Commun. Math. Phys. **110** 661-683.

[63] —— (1989). The scaling limit of self-avoiding random walk in high dimensions. Annals of Prob. **17** 91-107.

[64] Sokal, A. D. and Thomas, L. E. (1989) Lower bounds on the autocorrelation time of a reversible Markov chain. J. Stat. Phys. **54** 797-824.

[65] Spitzer, F. (1976). *Principles of Random Walk*. Springer-Verlag.

[66] Stein, E. and G. Weiss (1971). *Introduction to Fourier Analysis on Euclidean Spaces*. Princeton University.

[67] Stöhr, A. (1949-50). Uber einige lineare partielle Differenzengleichungen mit konstanten Koeffizienten III. Math. Nachr. **3** 330-357.

[68] Stoll, A. (1989). Invariance principles for Brownian local time and polymer measures. Math. Scand. **64** 133-160.

[69] Symanzik, K. (1969). Euclidean quantum field theory. Appendix by S. R. S. Varadhan. *Local Quantum Theory* (R. Jost ed.), Academic Press.

[70] Vicsek, T. (1989). *Fractal Growth Phenomena*. World Scientific.

[71] Wall, F. and R. White (1976). Macromolecular configurations simulated by random walks with limited numbers of non-self-intersections. J. Chem. Phys. **65** 808-812.

[72] Westwater, M. J. (1980). On Edwards' model for long polymer chains. Commun. Math. Phys. **72** 131-174.

[73] Witten, T. and L. Sander (1981). Diffusion-limited aggregation, a kinetic critical phenomenon. Phys. Rev. Lett. **47** 1400-1403.

Index

Appendix A

Recent Results

In this addendum I would like to summarize a few results that have been proved since the first printing of this book. I will only discuss some results directly relevant to the last four chapters of the book.

The method of "slowly recurrent sets" was used in [A3] to improve the estimate on $f(n)$ in four dimensions as discussed in Section 4.4. A subset A of Z^d is called slowly recurrent if it is recurrent, but $P(V_n) \to 0$, where V_n is defined as in the proof of Theorem 2.2.5. (By Theorem 2.2.5, A is recurrent if and only if $\sum P(V_n) = \infty$.) An example of a slowly recurrent set is the path of a simple random walk in four dimensions. In [A3] it is shown that there is a constant c such that

$$f(n) \sim c(\ln n)^{-1/2}.$$

Zhou independently gave an argument to show that

$$f(n) \asymp (\ln n)^{-1/2},$$

and Albeverio and Zhou [A1] also have proved the corresponding result for Brownian motions in four dimensions.

The equivalence of Brownian motion and random walk exponents in two and three dimensions (see Section 5.3) was extended to mean zero, finite variance random walks in [A2]. In the case of simple random walk, there has been some improvement on the rate of convergence to the intersection exponent. Let $b(r) = b(r, x, -x)$ where $b(r, x, -x)$ is defined as in Section 5.2 and $|x| = 1$. It has been shown [A5] that

$$b(r) \asymp r^{-\xi} = r^{-2\zeta}.$$

Also, for simple random walk [A6],

$$f(n) \asymp n^{-\zeta}.$$

Some estimates for the disconnection exponent (Section 5.5) were derived in [A10] and in [A7] it was shown that the simple random walk disconnection exponent, which is defined in a natural manner, is the same as the Brownian motion exponent.

For a detailed treatment of results on self-avoiding walks discussed in Chapter 6, I recommend the recent book of Madras and Slade [A8]. One interesting result that has come out since their book is a result of Toth [A9] showing that the mean square displacement of a bond "true" or "myopic" random walk in one dimension does grow like $n^{4/3}$, which is the conjectured for the (site) myopic random walk (see Section 6.5). While the bond walk is technically easier to handle than the site walk, there is no reason to believe that they should have different critical exponents.

The method of slowly recurrent sets was used to prove the conjecture about four dimensional loop-erased self-avoiding walk discussed in Section 7.7. It has been proved [A4] that

$$a_n \asymp (\ln n)^{-1/3},$$

where a_n is the normalization constant in Theorem 7.7.6 (b). In other words, the mean square displacement of the the walk grows like $n(\ln n)^{1/3}$.

Additional References

[A1] Albeverio, S. and Zhou, X., Intersection properties of Brownian motions in four dimensions, preprint.

[A2] Cranston, M. and Mountford T. (1991). An extension of a result of Burdzy and Lawler, Probab. Th. and Rel. Fields **89**, 487-502.

[A3] Lawler, G. (1992). Escape probabilities for slowly recurrent sets, Probab. Th. and Rel. Fields **94**, 91-117.

[A4] Lawler, G., The logarithmic correction for loop-erased walk in four dimensions, preprint.

[A5] Lawler, G., Hausdorff dimension of cut points for Brownian motion, preprint.

[A6] Lawler, G., Cut times for simple random walk, preprint.

[A7] Lawler, G. and Puckette, E., The disconnection exponent for simple random walk, preprint.

[A8] Madras, N. and Slade, G. (1993). *The Self-Avoiding Walk*, Birkhäuser-Boston.

[A9] Toth, B., The 'true' self-avoiding walk with bond repulsion on Z: limit theorems, preprint.

[A10] Werner, W., An upper bound to the disconnection exponent for two-dimensional Brownian motion, preprint.

Probability and Its Applications

Editors

Professor Thomas M. Liggett
Department of Mathematics
University of California
Los Angeles, CA 90024-1555

Professor Charles Newman
Courant Institute of
Mathematical Sciences
251 Mercer Street
New York, NY 10012

Professor Loren Pitt
Department of Mathematics
University of Virginia
Charlottesville, VA 22903-3199

Progress and Its Applications includes all aspects of probability theory and stochastic processes, as well as their connections with and applications to other areas such as mathematical statistics and statistical physics. The series will publish research-level monographs and advanced graduate textbooks in all of these areas. It acts as a companion series to Progress in Probability, a context for conference proceedings, seminars, and workshops.

We encourage preparation of manuscripts in some form of TeX for delivery in camera-ready copy, which leads to rapid publication, or in electronic form for interfacing with laser printers or typesetters.

Proposals should be sent directly to the editors or to:
Birkhäuser Boston, 675 Massachusetts Avenue, Cambridge, MA 02139, U.S.A.